NUNC COGNOSCO EX PARTE

TRENT UNIVERSITY
LIBRARY

BOWER-BIRDS

Oxford University Press, Amen House, London E.C.4
GLASGOW NEW YORK TORONTO MELBOURNE WELLINGTON
BOMBAY CALCUTTA MADRAS KARACHI CAPE TOWN IBADAN
Geoffrey Cumberlege, Publisher to the University

BLUE MALE SATIN-BIRDS AT BOWER

When the owner (*left*) goes away to feed, a rival (*right*) comes stealthily through the undergrowth and begins to wreck the bower. The marauder will also steal the coloured display-things (snail shells, blue feathers, greenish-yellow flowers) if he has time. He always flees when the owner swoops back (pp. 62 *et seq.*).

BOWER-BIRDS

THEIR DISPLAYS AND BREEDING CYCLES

A Preliminary Statement

BY

A. J. MARSHALL

READER IN ZOOLOGY AND
COMPARATIVE ANATOMY IN THE
UNIVERSITY OF LONDON

OXFORD
AT THE CLARENDON PRESS
1954

'Only fools will think it commends them to the English reader to decorate incongruously with such bower-birds' treasures as *au pied de la lettre, à merveille, bien entendu, les convenances, coûte que coûte, quand même, dernier ressort*.... Every writer ... who suspects himself of the bower-bird instinct should make and use some ... classification system, and remember that acquisitiveness and indiscriminate display are pleasing to contemplate only in birds and savages and children.'

A Dictionary of Modern English Usage
by H. W. FOWLER, Oxford, 1949

PRINTED IN GREAT BRITAIN
AT THE UNIVERSITY PRESS, OXFORD
BY CHARLES BATEY, PRINTER TO THE UNIVERSITY

PREFACE

THOSE who enjoy reading about strange birds, but who are not very interested in the internal processes that influence their behaviour, may wish to skip Chapters 2 and 3, at least for the time being.

The sub-title 'A preliminary statement' is used because, although this book is the result of more than twenty years' intermittent work with bower-birds, and summarizes the meagre available knowledge about all known species, it is, in fact, no more than a basis for future research.

Some of my conclusions are of necessity based on the study of far too few laboratory specimens. If one works (for example) on the mechanism of inheritance in insects or mice, or neuro-muscular transmission in squids, or on the blood chemistry of sea-squirts, one can easily sacrifice without compunction as many animals as are necessary to ensure that, within the limits of one's ability and apparatus, accurate conclusions can be drawn. But it is difficult, even if one felt so inclined, to go about a country killing statistically relevant numbers of bower-birds during each changing phase of their display cycles. The habitats of some of them are separated as widely as London and Moscow. Some of the organs reported on were secured, almost by lucky chance, at odd times in remote places during the war when they were crudely preserved in gin or whisky. (This sacrifice was not as dreadful as might be imagined: a very small amount of whisky will, in this respect, go a very long way.)

It is a pleasure to thank various organizations and kind people for assistance in past years. Much of the early work was done when I was a Grainger Post-graduate Scholar in Physiology in St. Paul's College, University of Sydney, and the recipient of a research grant from the then newly formed Association for the Study of Animal Behaviour. Post-war work on bower-birds has been carried out while I was a Beit Memorial Fellow, engaged primarily in the study of other problems in the Department of Zoology and Comparative Anatomy, Oxford, and later, with the help of a grant from the Royal Society, in the Department of Zoology and Comparative Anatomy, St. Bartholomew's Medical College in the University of London. At various times I have had the benefit of stimulating discussions with Professor Frank Cotton (Sydney), Dr. John R. Baker (Oxford), and the late Dr. F. H. A. Marshall (Cambridge) on various aspects of the study. Dr. N. Tinbergen (Oxford) has been good enough to read parts of the manuscript of the present book and I am indebted to him for several points of criticism.

I am especially grateful to Mr. G. Valder, who secured bower-birds for my aviaries, and to the Trustees of Taronga Park, Sydney, and the

Council of the Zoological Society of London, who gave them temporary lodgings at various times. Other assistance has been given me by the directors and individual members of staff of the Australian, National, Queensland, South Australian, West Australian, American, and Humboldt (Berlin) Museums of natural history and, in particular, by the staff of the British Museum of Natural History and the library staff of the Zoological Society of London. The editors of the *Australian Zoologist, Emu, Proc. Zool. Soc. Lond., Proc. Linn. Soc. Lond., Ibis, Nature, Wilson Bulletin (U.S.)* and *Biological Reviews* have allowed me to use material and illustrations that have appeared in their journals. I would like also to record my appreciation of the help given me by the staff of the Clarendon Press.

Many of the pictures are not my own. The names of gentlemen who have been generous in lending me photographs are mentioned beneath their respective contributions. The photomicrographic illustrations were made by my friends Mr. P. D. Mumby and Miss Olive Wilkinson. Finally, the many unacknowledged drawings which enliven the text are the work of my wife, Jane Marshall, who has helped in numerous other ways as well.

A. J. M.

LONDON
Autumn 1953

CONTENTS

MAP OF AUSTRALIA	viii
LIST OF PLATES	ix
MAP OF NEW GUINEA	x
1. INTRODUCTION	1
2. INTERNAL EVENTS IN THE AVIAN BREEDING CYCLE	7
3. THE INTERACTION OF EXTERNAL AND INTERNAL FACTORS IN THE AVIAN BREEDING CYCLE	13
4. SATIN BOWER-BIRD (*Ptilonorhynchus violaceus*)	26
5. SPOTTED BOWER-BIRD (*Chlamydera maculata*)	72
6. GREAT GREY BOWER-BIRD (*C. nuchalis*)	89
7. FAWN-BREASTED BOWER-BIRD (*C. cerviniventris*)	100
8. YELLOW-BREASTED BOWER-BIRD (*C. lauterbachi*)	106
9. REGENT BOWER-BIRD (*Sericulus chrysocephalus*)	109
10. NEW GUINEA REGENT BOWER-BIRD (*S. (X) bakeri*)	119
11. GOLDEN-BIRD (*S. (X) aureus*)	120
12. BROWN GARDENER (*Amblyornis inornatus*)	123
13. ORANGE-CRESTED STRIPED GARDENER (*A. subalaris*)	127
14. YELLOW-CRESTED GARDENER (*A. macgregoriae*)	130
15. GOLD-MANED GARDENER (*A. flavifrons*)	134
16. QUEENSLAND GARDENER (*Prionodura newtoniana*)	135
17. GOLD-CRESTED BLACK BOWER-BIRD (*Archboldia papuensis*)	141
18. CRESTED BIRD-OF-PARADISE (CRESTED 'GOLDEN-BIRD') (*Cnemophilus macgregorii*)	144
19. GREEN CAT-BIRD (*Ailuroedus crassirostris*)	148
20. WHITE-THROATED CAT-BIRD (*A. buccoides*)	152
21. STAGEMAKER (*Scenopoeetes dentirostris*)	154
22. THE EVOLUTION OF BOWER-BUILDING	165
23. ADDENDA	189
REFERENCES	194
SCIENTIFIC NAMES OF ANIMALS MENTIONED IN TEXT	201
INDEX	203
PLATES	at end

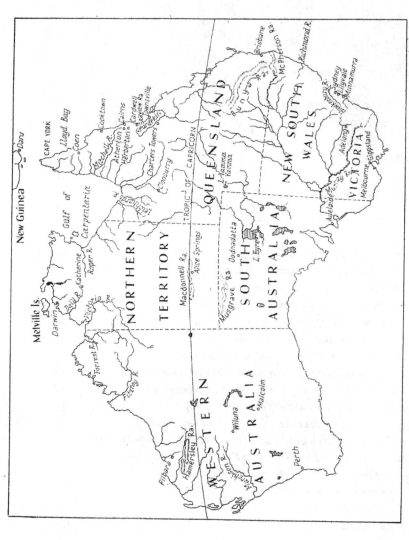

AUSTRALIA, SHOWING GEOGRAPHICAL RELATIONSHIP WITH NEW GUINEA AND IMPORTANT LOCALITIES MENTIONED IN TEXT

LIST OF PLATES

1. Blue male Satin Bower-birds at bower
2. Part of the internal mechanism of display and reproduction in birds
3. Part of the internal mechanism of display and reproduction (*continued*)
4. Home of the Satin Bower-bird
5. Satin Bower-birds in display at bower
6. *Above*: Display arena of young Satin Bower-birds
   Below: Female Satin Bower-bird at nest
7. Seasonal gonad states of the Satin Bower-bird
8. Seasonal gonad states of the Satin Bower-bird (*continued*)
9. Seasonal internal gonad changes in the Satin Bower-bird
10. Spotted Bower-bird and its bower
11. Display in the Spotted Bower-bird
12. The home, and the nest and eggs, of the Spotted Bower-bird
13. Seasonal internal gonad changes in Spotted and Great Grey Bower-birds
14. Bowers of the Great Grey Bower-bird
15. The home, the bower, and decorations of the Fawn-breasted Bower-bird
16. Bowers of the Yellow-breasted Bower-bird
17. *Above*: Female Regent Bower-bird at nest
    Below: Bower of the Regent Bower-bird
18. *Kunai*-grass, and upland rain-forest of New Guinea
19. Bower 'hut' and garden of the Brown Gardener
20. Victorian and Edwardian artists' impressions of the bowers of Gardeners
21. *Above*: Edwardian artist's impression of bower of the Yellow-crested Gardener
    Below: Modern photographic reality
22. Rain-forest of tropical north-east Queensland
23. Bower of the Queensland Gardener
24. *Above*: Green Cat-bird at nest
    Below: Nest and eggs of the Green Cat-bird
25. *Above*: Stagemaker, or Tooth-billed Cat-bird
    Below: Leaf-strewn stage in the rain-forest
26. Seasonal internal gonad changes in the Stagemaker

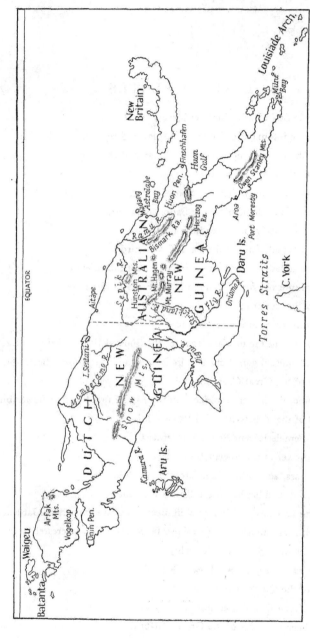

NEW GUINEA, SHOWING GEOGRAPHICAL RELATIONSHIP WITH AUSTRALIA AND IMPORTANT LOCALITIES MENTIONED IN TEXT

1

INTRODUCTION

THE bower-birds and cat-birds (Ptilonorhynchidae) are a colourful family of passerine or perching birds between 8 and 15 inches long which are confined to Australia, New Guinea, and nearby islands. It has been common knowledge for a century that the bower-birds build display-grounds and decorate them with a collection of objects that are used in display. The bower of each species is built on the ground. It has no direct connexion with the nest which is made in a tree, often several hundreds of yards away, much later in the season. In comparatively recent years it has become known that members of three genera paint the walls of their bowers with fruit-pulp, or wet powdered charcoal, or comminuted dry grass mixed with saliva. One species, the Satin Bower-bird, manufactures a tool to help it apply the paint. Another species, the Queensland Gardener Bower-bird, builds a bower which may reach a height of about 9 feet. This it decorates with living orchids. A gardener bower-bird of New Guinea builds a hut, plants before it a 'meadow of moss' and on this, with high discrimination, it arranges coloured fruits, flowers, fungi, and other objects that it gathers in the rain-forest. In Australia three species are known to enter human dwellings to remove coloured or otherwise distinctive decorations for their bowers. On the display-ground of the Spotted Bower-bird, silver coins, jewellery, and stolen car keys are occasionally found. Most of the species whose habits have been even briefly described are known to be quite exceptional mimics of other birds. Furthermore, their noisy and spectacular bower displays often take place outside the breeding season.

These and other singular attributes have caused a voluminous popular literature to spring up about the family. Much of this is nonsense. Most of it has been marred by anthropomorphic generalization, and all of it is unsupported by experimental evidence. In this literature (and in scientific journals) it has generally been claimed that sexual motivation plays little or no part in bower-building and allied phenomena, and that the principal motive for them is a consciously aesthetic one. It is often asserted that bower-birds carry out their extraordinary activities essentially for the sake of 'recreation'. It is generally suggested that they 'play' at 'playgrounds' or 'theatres' rather than display. The few ornithologists who have seen the birds in the bush have almost always credited them with intelligence (without experimentation), and one writer in a scientific

INTRODUCTION

periodical has seriously suggested that the Satin Bower-bird has the capacity for deductive reasoning. Apart from purely descriptive matter, little that has been written about the behaviour of wild bower-birds will stand up to critical inquiry. Even today the display-grounds of bower-birds are 'one of the strangest and least understood phenomena in nature'.[266]

The present contribution arises from a series of histo-physiological studies[162-3, 170, 175, 177] designed to test the validity of the 'recreation' hypothesis. By means of an examination of the reproductive cycle, ecology, and literature relating to each species, an attempt has been made to reach some generalization concerning the nature and function of bower-building and display, and, if possible, to describe these and associated phenomena in terms of animal rather than human behaviour. As a result of this, the following thesis is presented: as suggested briefly by Gould[104] and Darwin,[64] the bowers, display-grounds, and the activities that take place near them are primarily connected with sexual reproduction. These complex and remarkable phenomena are probably expressions of innate behaviour patterns that are annually called into play by the secretion of sex hormones during the period of pre-nuptial testis modification which begins at approximately the same time each year except in birds that inhabit dry areas of spasmodic and uncertain rainfall. At his display-ground, increasingly as the sexual season heightens, the male displays energetically and noisily, each according to his kind. The display-ground is the focal point of his territory, and to it is drawn a female which, in some species, watches intently but impassively as he performs. The display, which may continue for months, attracts and retains the female's interest, keeps off rivals, and helps synchronize the male and female reproductive processes until, when the environment becomes seasonally appropriate for nesting and the rearing of young, fertilization occurs and the female goes off to build her nest and to rear her young without masculine assistance. The male (in at least one species studied in detail) continues to display at the bower. It is not yet known whether bower-birds are polygamous. In any case, bower-building and display heighten reproductive efficiency, and are held to have arisen through the agency of natural selection. The theories of Australian naturalists that bower-birds are especially intelligent and that their display activities are largely 'relaxative', consciously aesthetic, and unconnected with the sexual drive are rejected, though of course it is not suggested that the birds do not enjoy the fantastic activities that they perform.

A few remarks are necessary concerning affinities and classification. It seems generally agreed nowadays that bower-birds are not as closely related to the birds-of-paradise as was once thought. According to

Stonor,[266] who made an anatomical appreciation of the family, they are clearly separate from all other birds as a 'singularly complete and isolated family of the acromyodian passerine birds which show no special relationship to any other'. Stonor recognizes the following genera: *Ailuroedus, Amblyornis, Chlamydera, Cnemophilus, Prionodura, Ptilonorhynchus, Scenopoeetes, Sericulus, Xanthomelus*. Since then a new genus and species, *Archboldia papuensis*, has been described.[228] More recently still, Mayr and Jennings[191] have questioned the validity of the New Guinea *Xanthomelus* as a separate genus. This doubt appears to be well justified, and so *Xanthomelus* has been combined with the Australian *Sericulus* in the present work.

As regards specific recognition, the views of Mayr and Jennings have been followed. Thus the Australian Cat-birds *Ailuroedus crassirostris* and *A. maculosus* are treated as a single species. Also, despite the appreciable physical differences between them, and the possible specific validity of *Chlamydera maculata* and *C. guttata*, these Australian bower-birds have been combined. It is regretted that so little is yet known of the display phenomena of the western *C. m. guttata*, for it would be nowadays generally admitted that a knowledge of the ecology of species is of value in the determination of taxonomic relationships.

Some bower-birds undoubtedly appear to show 'little inter-related connections' upon examination of preserved plumage, beaks, and limbs, but it is nevertheless surprising to have a modern opinion that 'bower-making has developed coincidentally or perhaps independently' among them.[126] When we consider the astonishing similarity of the display phenomena of, for example, members of allegedly comparatively unrelated genera *Ptilonorhynchus* and *Chlamydera*, it appears more than probable that their relationship is far closer than mere surface anatomy suggests. I propose to show that, diverse as some of the Ptilonorhynchidae appear at first sight, and however discontinuous is their distribution, they all (excluding one monospecific genus *Archboldia* of whose bower only fragmentary knowledge is available) appear to fall into three distinct groups: (1) the Stagemaker and two other cat-birds, (2) the avenue-builders (Fig. 1), and (3) what I will call maypole-builders (Fig. 2). In the light of this, and in view of the present uncertainty of some of the affinities, we must await impatiently the discovery of the still undescribed bowers of three more species in the upland rain-forests of New Guinea. I believe that a bower-bird's display-ground is just as characteristic as is its plumage, and that, at the present time, the description of its bower and display-paraphernalia is next in importance to the collection, preservation, and tentative classification of the initial skin.

The poetic family name 'bower-bird' was probably coined by the

←— *Sericulus*

  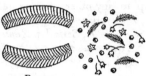

Platform　　　Sub-bower　　　　Bower

Ptilonorhynchus

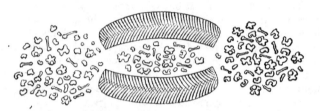

Chlamydera maculata

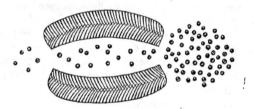

Chlamydera cerviniventris

Chlamydera lauterbachi —→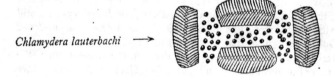

FIG. 1. Some avenue-type bowers and display-grounds.

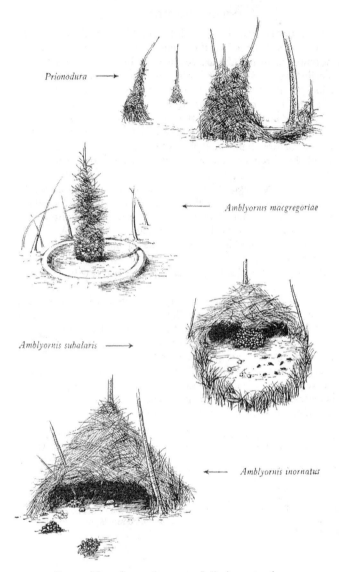

FIG. 2. Maypole-type bowers and display-grounds.

nineteenth-century English naturalist John Gould.[101] On his return from Australia, Gould announced to the Zoological Society of London, in 1840, that bower-birds had the extraordinary habit of building what the colonists called 'runs'. Of bower-birds' runs John Gould said: 'These constructions are perfectly anomalous in the architecture of birds. . . . [They] . . . consist of a collection of pieces of stick or grass, formed into a bower. . . . One of them (that of the *Chlamydera*) might be called an avenue. . . . They are used by the males to attract the females.' Fifty-three years later, Alfred Newton[204] observed: 'this statement, marvellous as it seemed, has been proved by many subsequent observers to be strictly true.'

Gould's information, repeated and extended in his later works, caught the imagination of the Victorian public. Very soon people were applying affectionately the expression 'bower-bird'* to others who indiscriminately accumulated possessions, including colourful rubbish. A modern and acidulous variant of this concept occurs in the course of one of Fowler's strictures (p. iv).

How erroneous and misleading are such comparisons will emerge in the following pages.

In the present account, the behavioural characteristics, sexual cycle, and plumage pattern of each species will be discussed in relation to each other and to the environment in so far as they can be in the meagre state of our present knowledge. The descriptions given will be brief and general. Lengthy details concerning plumage and other aspects of surface anatomy are to be found in Iredale[126] and in other accounts. Here I will provide merely a general physical impression of each species in relation to its other attributes and one which, if need be, should prove an adequate key for specific, but not sub-specific, recognition.

* There is a suggestion, based on probably inaccurate recollection, that the expression was current in late Georgian times. Thus, Trelawny, in his *Memoirs of Shelley* (1878), reports himself as saying, not later than 1822, that, 'You two have built your nest after the fashion of the Australian bower-birds.' In 1822 Australia was still known generally and officially as New Holland. Although Matthew Flinders had applied the name Australia (used previously in other connexions) specifically to New Holland as early as 1804, Sir Joseph Banks and other important personages frowned on such usage and it was not until the mid-twenties that the public began to follow the lead of Flinders and Governor Lachlan Macquarie. The Colonial Office continued to use the original name until at least 1849.[8]

2

INTERNAL EVENTS IN THE AVIAN BREEDING CYCLE

BEFORE discussing individual species of bower-birds and their attributes it is necessary to outline the basic events, internal and external, of the reproductive cycle in so far as they are thought to be understood. A bird, like all other vertebrate animals, is equipped with a system of sometimes highly developed sense organs which enable it to appreciate significant events in its environment. Thus, the effects of visual, auditory, tactile, thermic, olfactory, and other impressions impinge upon the central nervous system of each bird. Although individuals of a given species differ among themselves to some minor degree, and have the capacity to increase their chances of survival by learning, they lack the highly organized cerebral cortex that is found in the forebrain of the Mammalia and are, therefore, as far as we know, incapable of thought in the generally accepted sense of the expression. The central nervous system receives messages from the environment concerning danger, safety, hunger, repletion, warmth, cold, the actions and reactions of competitors, predators, its mate, other members of the flock, general well-being, and so on. In ways that are still not fully understood, such stimuli lead to further nervous, and endocrine, reactions, which in turn cause the individual to act in accordance with innate ('instinctive') behaviour patterns traditional to the species and to a given situation.

It now seems likely that it is a *combination* of external stimuli which, acting in unison with the internal reproductive rhythm, leads to the seasonal maturation of the sex organs of various species and makes it possible for mating and reproduction to occur at the appropriate breeding season.[164-5, 171, 180, 278]

The ancients knew that breeding seasons occurred, but we still do not know the precise combination of factors that initiates the reproduction of even a London sparrow. This much only (see Fig. 3) seems probable: At approximately the same time each year, varying according to the species, a combination of environmental factors influences the central nervous system. In turn the hypothalamus releases a stimulant which, travelling down the hypophysial portal system, activates the *pars distalis* of the anterior pituitary gland (Pl. 2 *a*). This organ then liberates hormones which flow through the blood-stream to the reproductive and

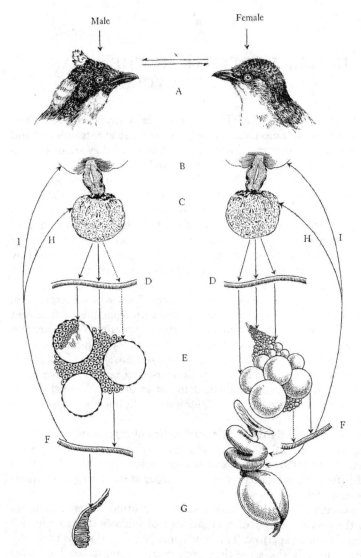

Fig. 3. Some of the suspected mechanism of the reproductive cycle in birds. (Partly after Romanoff and Romanoff.) The head-figures depict the Queensland Gardener Bower-bird (*Prionodura newtoniana*).

(*For Explanation see opposite page.*)

INTERNAL EVENTS IN THE AVIAN BREEDING CYCLE

other organs of the male or female as the case may be. The pituitary gonadotrophins stimulate the sex organs to activity, resulting in the liberation of male or female sex hormones which in turn cause the seasonal development of other organs that are accessory to reproduction. These hormones reach the brain, and there call into operation behaviour patterns concerned with combat, sexual display, the discovery and acquisition of a mate, nest-building, and finally coition and actual reproduction. Such activities proceed in a more or less orderly manner according to the state of the weather, food supply, and other vital conditions of the environment.[165] But however favourable the environment for reproduction,

---

Explanation of Fig. 3

BOTH SEXES

External stimuli (A) from the environment (including the opposite sex) are received by the exteroceptor organs. A chemotransmitter is probably liberated from the hypothalamus (B) at the base of the brain and passes down the hypophysial portal system to the anterior pituitary (C). (See also Pl. 2.) As a result of these and other events, the anterior pituitary begins a seasonal increase in activity.

MALE

A tubule-ripening hormone is liberated from the anterior pituitary (C) and flows into the blood-stream (D). There is also produced an interstitial cell-stimulating hormone which causes the interstitial (Leydig) cells of the testis (E) to produce a male sex hormone (Testosterone) which flows through the blood-stream (F) and modifies the accessory sexual organs (e.g. G). It flows also (H) to the anterior pituitary where it has a regulating effect and (I) to the brain where it probably calls into play innate behaviour-patterns concerned with display, mating, and reproduction. It is possible too, that a third, Prolactin-like hormone may be liberated from the anterior pituitary (dotted line) and may have at least an indirect effect on the testis (E) in connexion with the post-nuptial metamorphosis (see Pl. 2, 3) at the end of the display-season.

FEMALE

A follicle-ripening hormone is liberated from the anterior pituitary (C) and flows through the blood-stream (D) to the ovary (E). There is also probably produced a luteinizing hormone. The above hormones are probably essential for the development of the follicle and the egg. Prolactin is next liberated; it depresses the production of the two former hormones and causes broodiness to occur. Meanwhile a female sex hormone (Oestrogen) has been liberated from the ovary in increased quantities: it flows through the blood-stream (F) and reaches and seasonally activates the accessory sexual organs (e.g. G). It is possible that a Progestin-like substance is also liberated from the ovary (short dotted arrow). Meanwhile, Oestrogen flows (H) to the anterior pituitary and probably has an inhibiting action on the production of F.R.H. It flows also (I) to the brain where, in association with other hormones, it probably calls into play behaviour-patterns concerned with mating and reproduction.

N.B. It must be emphasized that some of the above reactions should still be considered hypothetical.

there is a period which follows breeding in which it is impossible for another sexual cycle to be started. This has been called the refractory period.[24-25] Thus, after birds produce their brood, or broods, in spring and early summer, the reproductive apparatus of the male becomes unresponsive to further stimuli. This phase lasts until late summer, autumn, or even longer. Thus there is an internal brake on breeding activity.

There is at present considerable controversy concerning the precise nature of the refractory period.[7, 21, 32, 169, 194] Bissonette and Wadlund,[24] in the course of photo-stimulation experiments with captive Starlings, demonstrated that there was a 'failure of the sex mechanism to maintain a high rate of germ-cell multiplication and maturation beyond a certain period'. The more stimulating the effect of the artificial illumination, the more quickly is the mechanism thrown out of gear. Later[232-3] it was found experimentally that the testes of spring-breeding House Sparrows were unresponsive to photo-stimulation as late as September. By November, however, the reproductive apparatus of the Sparrows had reached a condition sufficiently responsive to photo-stimulation to allow spermatogenesis to occur.

The testis cycle and the refractory period

The apparent explanation of the refractory period in so far as the testes are concerned is as follows:[166, 169] In the young bird the inactive tubule contents are free of lipids (Pl. 2 b). The interstitium contains juvenile Leydig cells that exhibit cholesterol-positive lipid globules from which are probably manufactured the male sex hormone. These globules increase in quantity until many of the cells reach an almost maximum size at a time when the contents of the adjacent spermatogenetic tubules (and the bird's plumage in some species) also indicate approaching sexual maturity. In the sexually quiescent adult the interstitium generally consists for the most part of comparatively small lipid Leydig cells. These contain fuchsinophil elements which are readily seen after the soluble lipids are removed by embedding in wax. As the sexual season heightens, the Leydig cells increase in size and lipid content. Varying in quantity and arrangement from species to species, considerable aggregations of heavily lipoidal Leydig cells can now be seen (Pl. 2 c). The ripening spermatogenetic products expand the seminiferous tubules and this causes a wide dispersion of the clustered Leydig cells so that few are found in any given microscopic section. At the height of spermatogenesis, and the beginning of lipoidal metamorphosis of the germinal epithelium, the Leydig cells become almost exhausted of their lipid content (Pl. 2 d). An increase in the number of mitochondria can be seen in their cytoplasm. Soon a fuchsinophil cell appears in which it is not possible to demonstrate the presence of cholesterol. There is a difference of opinion concerning the relative function of the lipid and the fuchsinophil cell. Sluiter and van Oordt[260-1] call the latter (which has also a vacuolated form) the 'secretory cell' and declare that it manufactures the sex hormone. I,[166] on the other hand, believe the lipid cell to be the site of hormone production and consider that the fuchsinophil cell is a lipid cell that is completing its cycle and is about to disintegrate or to give rise to another generation of juvenile Leydig cells.

Concurrently, when the tubules have reached their maximum size and after spermatozoa are shed, some tubules undergo a lipoidal metamorphosis (Pl. 3 e). (This is quite distinct from the lipophanerosis observed during vertebrate spermatogenesis.) Soon the contents of all tubules become almost wholly lipoidal (Pl. 3 f). The tubules rapidly collapse as the necrotic spermatozoa and other debris clear away. Meanwhile, a new generation of small juvenile Leydig cells arises in the spaces between the shrunken tubules (Pl. 3 f) and the interstitium is invaded by new blood-vessels. There now arises a prolific growth of fibroblasts which build up a new testis tunic (Pl. 3 g) inside the distorted and fragile old wall. The new Leydig cells become meagrely lipoidal and fuchsinophil; and eventually become sufficiently mature to respond to external stimuli via the anterior pituitary gland. The refractory period is ended. We shall see that the above cycle of events occurs in bower-birds, as also in all other species so far investigated.

The interstitial exhaustion and tubule steatogenesis are associated with the post-nuptial moult, and is probably responsible for the brief period of sexual quiescence that now ensues.

In many bird species a sustained period of autumnal display, song, and territorial behaviour begins as the new interstitium develops and the post-nuptial tubule lipids disintegrate. It is inconceivable that the relatively large reservoir of cholesterol-positive tubule lipids is functionless. I believe that it may have an endocrine function, but this has not been proved. The gradually disappearing cholesterol may be converted to androgens which, like testosterone from the Leydig cells, may be involved in a reciprocal relationship with the hypophysis as well as having possibly the function of preserving aggression in food-gathering, &c., during the period of seasonal interstitial immaturity.[180] The metamorphosis described above, in addition to the subsequent 'spring' development of interstitium and seminiferous tubules, constitutes a physical basis of the internal rhythm postulated by F. H. A. Marshall[183] and Baker.[10]

It is possible that the anterior pituitary gland, too, may undergo a refractory phase at the termination of the breeding season, but so far no wholly convincing evidence in support of this has appeared. It may be that the testis metamorphosis is caused by temporary hypophysial exhaustion and the consequent cessation or reduction of the output of pituitary gonadotrophins, or by other means (Fig. 3). It will be recalled that injections of prolactin, for example, lead to a swift and profound testis atrophy in pigeons,[142] cockerels,[202] and passerine birds.[179]

On the other hand it has been shown that among Arctic birds whose breeding cycle has been halted by the lack of stimuli favourable to breeding, or the presence of stimuli unfavourable to breeding, the testes of non-breeding individuals metamorphose before the completion of spermatogenesis. The testes of species that breed successfully do not metamorphose until much later when the young have appeared. I have suggested[174] that this premature tubule steatogenesis may be caused by a diminution, rather than a complete exhaustion, of hypophysial activity and that this results from a lack of the environmental stimuli that are necessary for successful reproduction in the species in question. If the Leydig cells are no longer stimulated by gonadotrophins it might be expected that they, and also the seminiferous tubules, would be halted in their activity. The experimental removal of the anterior pituitary has been shown to result in tubule metamorphosis and interstitial regeneration (Pl. 3 h) in the fowl's testis in a manner very similar to that obtaining in the Arctic non-breeder.[181]

Whether or not the hypophysis also undergoes a refractory period, there can be no doubt that the intervention of such a period in the testes, regardless of day-length and other environmental factors, causes a cessation of breeding and prevents sexual exhaustion, over-population, and resultant over-competition for the available food supply. In temperate-zone species, the refractory phase seems generally to last until the autumn, when breeding activities are inhibited by lower temperatures, lack of sunshine, food scarcity, lack of breeding cover, and possibly shortness of day. The phenomenon has not yet been studied in the tropics but it would seem that, if the refractory period were long enough, it would be an important internal factor in the timing of breeding seasons and migration. Once the refractory period is past, gonadotrophins from the anterior pituitary stimulate the primary sex organs and there begins the seasonal liberation of sex hormones. These cause the development of the accessory sex organs. Equally importantly, they flow to the central nervous system and call into play inherent behaviour patterns concerned with display and reproduction.

3

THE INTERACTION OF EXTERNAL AND INTERNAL FACTORS IN THE AVIAN BREEDING CYCLE

WE can now consider birds as free-living animals in touch with all of the varied stimuli that keep their breeding cycles in step with the sun and ensure for their young an arrival time suitable for their survival. In many birds the ending of the refractory period appears to signal the onset of a post-moult sexual recrudescence during the autumn months. Morley[198] has listed sixty-three British species which behave in this way and such autumnal singing and display has been widely recorded on the European continent, in North America, Australia, and elsewhere. We shall see that such post-nuptial display is especially pronounced in bower-birds. In some British species, notably the House Sparrow, Starling, Wood Pigeon, Stock Dove, Robin, and Blackbird, this characteristic post-moult activity may occasionally lead to autumn, or even winter, reproduction[287] if weather and food conditions remain propitious. Some of these occasional out-of-season breeders exhibit pronounced autumnal sexual behaviour in transported American stock.[1, 288]

Among the few British birds whose breeding cycle has been studied in some detail, the Starling begins its seasonal increase in testis size in September[31] and this is accompanied by pronounced sexual activity. Likewise, in autumn, the Mallard begins a similar resurgence, accompanied by mating behaviour.[116] The autumn song of the Robin has been admired in verse and story for centuries[140] and its endocrinological basis has recently been studied in some detail.[176] Autumnal song begins while the testes are still minute but when the adult interstitium begins to mature and when the interstitial cells of the young of the newly hatched spring brood show the first signs of cholesterol-positive lipids. Spermatogenesis, however, will not usually begin until early January, and reproduction will not occur until the following April. We shall later see certain similarities among the bower-birds.

In the Rook a similar post-nuptial period of testis reorganization has been demonstrated,[180] and when this ends the birds begin their autumn mating flights,[33, 54] refurbish their nests and a small proportion of the males achieve spermatogenesis. Breeding very occasionally occurs in

14   THE INTERACTION OF EXTERNAL AND INTERNAL

November,[271, 277] but the regular seasonal gametogenesis is inhibited until early in February.

We shall see in subsequent chapters that the four species of bower-birds and cat-birds whose sexual cycles have been investigated histologically all undergo a post-nuptial testis metamorphosis involving tubule breakdown and an accompanying interstitial development. So, it will be seen, there is nothing remarkable in the return of the birds to their territories, the building of bowers, and the performance of out-of-season displays during the post-nuptial phase of interstitial regeneration.

In many species of birds this premature nuptial activity does not continue because of the inhibitory action of winter conditions and, conversely, the absence of the external factors that traditionally stimulate the cycle to pronounced activity in early spring. These environmental inhibitors and stimuli, which will be discussed below, are potent factors in the timing of the breeding cycle of the various species. They prevent matings from occurring at times of year when the young would not survive; and they cause reproduction to occur at times most propitious for their survival. It is exceedingly noteworthy that at least one species of sea-bird, breeding on a tropical island surrounded by a probably inexhaustible food supply, reproduces about every 9·6 months, i.e. as soon as each post-nuptial refractory period comes to an end. This is the Sooty Tern of Ascension Island,[45] which is apparently exposed to none of the inhibitory influences that 'brake' the post-refractory development of temperate and sub-polar species.

We have seen above that there exists within the bird an internal reproductive rhythm embracing successive periods of high activity (culminating in reproduction) and reorganization and comparative quiescence (the refractory period). After this post-nuptial rehabilitation, the internal reproductive apparatus is once more susceptible to the external stimuli to which the species traditionally responds.

Environmental regulating factors

For centuries men thought that it was simply the rising temperature of spring that induced breeding and this view was generally held until comparatively recently. Again, some have held, even until the present day, that each animal has within it a mechanism, something like a natural alarm clock which, quite independently of external influences, operates annually to enable a given species to breed at a certain precise time every year. Thus, it was suggested[25] that the breeding cycle and migration in some birds might be governed by 'inherent rhythms of the anterior pituitary more or less fixed in the absence of, or without responsiveness to, external, usually stimulating, factors...'. About the same time, however, Baker[12] calculated that 'if in any plant or animal there were an internal rhythm controlling an annual cycle, and if this rhythm were incorrect even to the minute extent of 6 minutes in the year, then if it were breeding in the spring at the end of the last glacial epoch in the northern hemisphere it would be breeding at precisely the wrong season of the year (autumn) now!' From what is known of the time factor in biological reactions any such exactitude is inconceivable. It is certain, then, that each annually breeding organism must be influenced from time to time by external factors that act as a regulating 'finger' to keep the 'clock' in step with the seasons. If this were not true, breeding could occur at unpropitious times of the year and the young would freeze or starve. We have seen that odd individuals of various species do indeed breed at inappropriate times of the year. These rarely bring their young to maturity.

There seems no reason to doubt that during the stress of the slowly changing climate and topography of the earth's surface over the centuries, natural selection has operated and compelled the evolution of breeding seasons. The individuals whose internal apparatus achieved the hereditary capacity to respond to external factors which ensured spring-born young would pass on the capacity to unborn generations because young born in the spring would survive. In turn, these would successfully reproduce their kind in the benign weather and profuse food supply of spring. The extraordinary malleability of the animal organism in its power of internal adjustment to the environment is shown in numerous birds and other animals. Thus of two particular Antarctic penguins, the Emperor brings out its chick in the darkest, coldest, and most tempestuous period of the year and the Adelie, living in the same latitude, hatches its egg during the lightest and warmest month. Levick[147] has described the astonishing differential growth-rate of the two species—the 'mushroom-like' development of the young Adelie, which enables it quickly to mature sufficiently to survive the following winter, contrasted with the slow growth of the Emperor chick which is carried and warmed on the feet of the parents until the arrival of better weather. Among mammals, Stieve[264] has collected fascinating examples of differential intra-uterine developmental rates which have been recorded even within the same species. Again, among certain European and North American bats another interesting mechanism, seemingly an adaptation to the northern winter, has been described.[13] These animals copulate in the autumn and the uterine horns of the females become distended with spermatozoa. But ovulation, and, it follows, conception, do not take place until the following spring and, as a result, parturition cannot occur until the mild weather and abundant food harvest of the following summer.

The intensive study of the environmental factors that stimulate and 'anchor' the breeding cycle began about 25 years ago with the epochal discovery by Rowan[243-4] of Alberta who brought captive male Junco Finches and later[245] American Crows to sexual and migratory activity merely by switching on added rations of electric light. Finches continued their response at temperatures of about 50 degrees below zero Fahrenheit. It was suggested that the naturally increasing day-lengths in spring are responsible for the annual sexual resurgence and migration. Rowan's work began an era of photo-experimentation. In laboratories throughout the world all kinds of animals were shown to be susceptible to photo-stimulation. For a time most people thought that light fluctuation would explain satisfactorily the whole problem of breeding seasons. A few people think so even today. Meanwhile, the Japanese Miyazaki[196] reported that for generations his people had exposed pet silvereyes to increased illumination (*yogai*) by means of candles in order to achieve unseasonal singing. About the same time it became generally known that Spaniards had for at least a century obtained increased egg-production by photo-stimulation of domestic fowls. This method has now become almost universal amongst poultry farmers. Meanwhile, Dutchmen[62] recalled the almost forgotten practice of putting cage-birds into the *muit*, or summer darkness, and then exposing them to artificial lighting in autumn in order to obtain winter song. Rowan[246] had meanwhile come to the conclusion that it is not light *qua* light that causes gonad development, and resulting migratory tendencies, singing, and associated reproductive phenomena. He came to believe that increased metabolism, allowed by the prolonged wakefulness caused by the increased daylength, is primarily responsible. Since then, the experiments which seemed to show that this was possible have been repeated by other people but substantiation of Rowan's latter theory has not been obtained. It would be logical to suppose that if light somehow stimulates the nervous system to cause hypophysial activity, then increased day-length after the winter solstice might indeed lead to the gonad recrudescence that initiates reproduction. But many mammals mate in the autumn and so drop their young in spring; and the experiments of Yeates[290] seem to have shown that prolonged *decrease* in daylength leads to mating in domestic sheep.

On the face of it there now seems to be overwhelming evidence that sexual periodicity in numerous species of temperate zone birds is somehow controlled by light and dark fluctuations. Such indeed may be the case, but an analysis of the experiments shows that most of them have been carried out under conditions that had little or nothing in common with the normal events of the lives of the birds involved. Most of the experiments have been crude in the extreme, and have involved merely the photo-stimulation of caged males for various daily periods with

FACTORS IN THE AVIAN BREEDING CYCLE

different intensities and wave-lengths. Before it can be unquestionably accepted that light fluctuation causes the pre-nuptial sexual recrudescence in nature, far more elegant experimental techniques than those used in the past must be devised. It is one thing to stimulate a bird's neuro-hormonal apparatus by massive illumination but another to show that the small difference in daylength naturally occurring between 21 December and 8 January does the same thing in, for example, the British Robin in the Midlands where the winter is so often very dull and foggy. Rowan was able to photo-stimulate finches to sexual maturity at temperatures of 50° below zero F. But that does not prove that temperature is unimportant in the timing of the sexual cycle under normal conditions.

Whether light fluctuations are initially influential or not, there now appears to be sufficient evidence to support the conclusion that, certainly after the pre-nuptial gametogenetic leap, the sexual cycle of a given pair of birds, or even a colony of birds, is governed by the week-to-week succession of environmental experiences, including the behavioural-interactions between the sexes. In short, the cycle is accelerated by external factors such as weather, food supply, sexual display, nest-site availability, and other factors to which the species is innately bound to respond. A lack of such factors, or the presence of certain inhibitors such as cold, hunger, and fear, halts or delays the cycle.

Numerous published data indicate that the breeding season fluctuates according to the mildness or severity of the weeks preceding nesting. When it became known that the British winter of 1946–7 was one of the coldest on record I thought it worthwhile to collect the gonads and stomach contents of a small series of four common Oxford passerines—the Robin, Chaffinch, Great Tit, and Blue Tit—in the middle of March in the hope that the following winter would be normal and the same species might be obtained in precisely the same wood exactly one year later.[165] The winter of 1947–8 was unusually mild: the proposed second collection was made. The experiment was designed to rule out daylength as a factor and to test the effects of weather fluctuation.

There proved to be a great disparity between the gonads taken after the 'hard' winter and those collected after the mild one. From 13 to 15 March 1947, none of the four species showed a greater development than primary spermatocytes, whereas those collected exactly one year later had all reached secondary spermatocyte stage, and three of the four species had developed spermatids or spermatozoa. Analysis of weather and breeding records showed that the hard winter was followed by an abnormally bright spell (after the collection date) and that the surviving birds of at least three species bred at about the normal time! It is, of course, extremely difficult to evaluate the individual importance of such factors as sunshine, temperature, and food abundance. Great care was taken to provide an adequate food supply for all birds collected in both winters. So, unless some special food component was operative, it seems that sunshine and temperature must have been particularly influential in the timing of the cycle.

A further proof against light in any form as the overall regulator is provided by the irregular cycles of many species of birds in countries where rainfall is scanty or spasmodic. Here, quite often, breeding dates are indubitably geared to rainfall or its immediate effects. There is evidence[38] that north-western Australian birds can be divided into several groups. Some (certain pigeons and quail and many passerines) breed at any time of the year following heavy rain. The seed-eating Budgerigar and Zebra Finch breed after the growing of grass following such rain. Certain species ovulate in a particular calendar month, but only if conditions are normal for the period. Finally, a few large birds-of-prey have a regular breeding time irrespective of environmental fluctuation. Spring-nesting West Australian species breed in autumn after heavy unseasonable rainfall. A little beyond the area of rainfall, there is no indication of unusual breeding.[253] The Dusky Wood-Swallow nests in the Western Australian autumn (when light is decreasing) if rain falls—and nests again in mid-winter if it rains again.[237] The central Queensland Plum-headed Finch lays only when heavy rain brings on a good crop of grass with seeds; the finches begin building immediately the grass becomes long. Scrub-turkeys in the same area begin raking rubbish on to their old incubators only after the first rains of the season. If the 'wet' continues they will lay. In one season of abnormal rainfall these megapodes laid in wintertime.[236] They rely, of course, on heat generated by the damp, decaying vegetation of their incubators to hatch the eggs. In south-western Queensland, where wet winters are exceptional, several kinds of spring-breeding birds have been observed to nest before the winter solstice after unseasonable rain.[156] In this respect evidence concerning Galapagos finches (*Geopspiza*) is of importance. Members of this genus apparently breed from mid-December to April during the wet season when in a wild state near the equator. In captivity in California, however, they generally nested from March to November, the breeding period of most North American birds.[213] If the captive equatorial finches had been native species their breeding during March–July would no doubt be ascribed to day-length. It would seem that some xerophilous or other species, which have of necessity evolved a neuro-endocrinal response to rain or its effects, will breed at any time of year after a short post-nuptial period of testis reorganization. A long, fixed period of reorganization might be fatal—the species would, perhaps, be refractory to external stimuli until long after the next rainfall and the passing of its beneficial effects. There seems little reason to doubt that various

FACTORS IN THE AVIAN BREEDING CYCLE   19

species (or even different populations of one species)[34] have evolved a breeding reaction to dissimilar environmental complexes and so enabled themselves to breed in all manner of habitats quite independently of the movements of the sun *per se*. In this respect the work of Wagner and Stresemann[278] is of outstanding importance.

An invaluable method of inquiry into the causes of breeding seasons—and one that has not been seriously exploited—is the examination of environmental factors, or lack of factors, that *inhibit* breeding. Spasmodic non-breeding in birds and other animals is extremely common in all habitats subject to violent environmental fluctuations. The complete inhibition of breeding of most species for 16 months in central-west Queensland has been reported.[18] It was noticed long ago that in some years many North Greenland birds fail to reproduce. The non-breeding of the Long-tailed Skua was correlated with the absence of its normal food, the lemming. Other observations in Novaya Zemlya[248] and in East Greenland[151] indicate that inhibition of breeding in the Long-tailed Skua, Pomatorhine Skua, and Snowy Owl is closely related to the lemming cycle which, as Elton[76a] has shown, occurs in a regular 3–4-year rhythm. When lemmings fail there is widespread non-breeding among lemming-feeders. On the bleak, foggy island of Jan Mayen (east of Greenland) we found[174] that non-breeding in some species appeared to be caused by the lack of nesting sites safe from fox-persecution. Certain species that show no great conservatism in site-selection elsewhere nested almost entirely on lofty cliffs and pinnacles on this fox-infested island. Passage waders touch down in thousands, but in 1947 only a few remained through the summer, and these did not breed. No wader has ever been recorded breeding on this island. Stomachs of the many waders shot by our expedition contained vegetable matter only. The testes of summer residents contained spermatozoa, so failure to breed may well have resulted from inability of the females to find the sort of food characteristic of the traditional nuptial area. Land and lake birds of many species have been reported from Jan Mayen by various expeditions, but very few breed there even though every day at the summer's peak provides 24 hours of light. Sunshine, however, rarely penetrates the heavy blanket of fog surrounding the island, and most migratory and dispersive species do not find there an environmental pattern appropriate to their traditional reproductive needs. The only birds that breed freely are cliff-nesting seafowl which can escape the fox population (which is confined to the island except during some winters when pack-ice allows egress across the sea to the Greenland coast) and can find unlimited food in the surrounding sea. The testes of non-breeding species underwent the same spermatogenetic cycle as those of the breeding species, but in development lagged behind those of paired breeding birds and often underwent premature metamorphosis. It seemed that both sexes of birds in a flock became potentially capable of reproduction but that lack of suitable external stimuli (e.g. those provided by a satisfactory breeding site and/or appropriate food) inhibited the final stages leading to fulfilment. Broadly comparable data have been obtained from a study of non-breeding Australian desert birds.[135a]

Tropical birds, too, present a strong argument against the overall control of the breeding cycle by light fluctuations.[197] We know now that most of the old travellers' tales of equatorial animals 'breeding all the year round' are untrue. Many tropical animals seem to have breeding seasons as sharp as those of the same families living in Tasmania, Sussex, or Vermont. The Oxford University Expedition[11] which established a laboratory for one year in the remarkably unvarying climate of the New Hebrides proved conclusively that the endemic race of the Golden Whistler has as sharp a breeding season as that of another race in southern Australia.[14] In the New Hebrides (below the equator) the species ovulates when the days are growing *shorter*. In southern Australia (below the equator) it ovulates when days are growing *longer*.

It could be suggested, of course, that the cycles of some of these little-known tropical birds are, like that of the Sooty Tern, not annually recurrent. It could be that in a comparatively unvarying equatorial climate, a cycle may culminate every eight or nine months without anybody suspecting that such was the case. No evidence is available to support such an idea. One tropical vertebrate has been studied over a period of years partly to investigate the matter. Thus, the Giant Fruit-bat of Ceylon, living at about lat. $7°$ N., has an exceedingly sharp breeding season. All females of a given colony become pregnant in a few weeks in November and December.[164] Eleven years after the first collections were made, bats taken in the same locality became pregnant at precisely the same time of year, despite the negligibility of light fluctuations, and the unusual stability of their hot, humid, near-equatorial rain-forest environment.[179] It is certain that internal rhythm (involving oestrous, pregnancy and lactation anoestrous) alone cannot account for this remarkably regular annual conception-date. If external stimuli of some kind were not involved, metabolic variation within the individuals would cause some irregularity over the years. We have no precise information concerning regular seasonal changes of the food supply, or other external stimuli which may influence fruit-bats as breeding regulators. If we are to fall back on the facile photo-periodicity hypothesis to explain such breeding regularity we must consider the following data: The days at the conception date (late November) have become only about one hour shorter than at the 'summer' solstice (21 June). When the animals come into oestrous the days are shortening at a rate of less than half a minute per day.

There has been no recent change from lengthening to shortening days. And what of the bats living precisely astride the equator?

In the case of birds, it is possible that a genetically fixed rate of post-nuptial testis rehabilitation may proceed to a certain threshold where any environmental stimuli (including behavioural inter-actions between the sexes) which influence the anterior pituitary can then be operative on the now again responsive gonads. Display, for example, has been described[267] in flocking Sharp-tailed Sandpipers in tropical Queensland (at lat. 15° 22′ S.) in March and early in April shortly before the birds' migratory departure north towards their Arctic breeding grounds. In the case of juveniles, a fixed maturation rate may be the essential timing factor: once a certain sexual maturity is reached, environmental factors may operate and so achieve a neurohormonal threshold sufficient to cause display, sexual synchronization with the flock, further androgen production and departure. When Emlen and Lorenz[77] implanted sex-hormone pellets in free-living, sexually inactive California Valley Quail, neighbouring untreated birds also showed accelerated mating behaviour.

Should it be considered that the concept of a semi-autonomous internal rhythm is improbable, it may be pointed out that a wholly autonomous one probably occurs in Man and other primates. Once the reproductive rhythm is initiated in the human female it seems to be continued automatically through the inter-action of the anterior pituitary and the ovary. The follicle-stimulating hormone of the pituitary is discharged into the blood-stream. It flows to the ovary where it stimulates the production of oestrogen. This female sex hormone is in turn discharged into the bloodstream. It induces the proliferation of uterine endometrium and causes other changes. When the oestrogen reaches a certain threshold it suppresses the discharge of follicle-stimulating hormone (from the pituitary) and so, in turn, oestrogen too is reduced in output. The luteinizing hormone is now liberated from the anterior pituitary. Ovulation occurs. The luteinizing hormone stimulates the development of the corpus luteum in the empty ovarian follicle that was formerly occupied by the egg. The new corpus luteum liberates progesterone. As the concentration of this hormone rises in the blood-stream the discharge of luteinizing hormone is checked. The corpus luteum begins to fail and endometrial sloughing occurs. Meanwhile, the oestrogen threshold has become low and the secretion of follicle stimulating hormone begins once more. So another monthly cycle begins.[23]

As everybody knows, various causes, quite apart from pregnancy, cause irregularity in the human cycle. Likewise, no internal avian rhythm, as has already been pointed out, could be expected to culminate at the appropriate reproductive time for many seasons in succession. The environment must operate to keep the cycle regular over the years.

The 'species requirement' for successful reproduction

After the initiation of the 'spring' pre-nuptial spermatogenesis (which even in temperate zones is now known often to occur in winter in several species) the male sexual cycle is hastened by sunshine, warmth, and food abundance and retarded by their absence. The males of all species studied far outstrip the females in gametogenetic development and achieve spermatogenesis within a few weeks. This leap in testis development is accompanied by a search for a mate, the taking up of territory and various specialized display activities, including song. The males of some species, including especially certain bower-birds, produce spermatozoa months before the females ovulate. Although the male has often been potentially ready to reproduce for many weeks, ovarian development is slow in all species so far studied and ovulation seems to be delayed until the changing environment presents to the female exteroceptors a pattern—a 'species requirement'—to which she has an innate capacity to respond. Thus, the phenomenon of 'Arctic non-breeding' and thus, too, the failure of desert and other species to breed in times of drought and their quick reproductive response after rainfall—irrespective of the movement of the sun. It is the external factors enabling the culmination, not the initiation, of the breeding cycle that ultimately 'time' the season.

We do not know what is this precise 'species requirement' for ovulation in even the commonest kinds of wild birds. The appearance in the environment of large quantities of the food on which the young are traditionally fed may be a highly important stimulus. As yet there are no experimental data to suggest that such an appearance actually stimulates the reproductive apparatus to activity, but there is accumulating a volume of evidence that makes it clear that most species breed at the height of periodic supply of the protein food on which they feed their young.[172] If this were not so the young would not survive. We will see that this applies particularly to certain of the bower-birds.[162]

Very young, small birds are inevitably at the mercy of their size and must perish if not almost constantly kept warm and well fed. The surface of an object increases as the square of its diameter, its volume as its cube. Thus the smaller a bird is, the greater is its surface in relation to its bulk. Birds, of course, are homoiothermous but at the same time most newly hatched young are naked and lack the sub-cutaneous insulating fat deposits that they later acquire. Therefore, although the nest-lining (when present), the proximity of their fellows, and the warmth of the parent (when present) make for heat-conservation, the relatively huge surface of a tiny chick means that the production of heat is a tremendous factor in its chance of survival because the loss of heat is proportional to the surface of the object from which it escapes. Heat is obtained from the oxidation of food. Therefore, to offset their relatively great heat-loss, small nestlings must consume enormous amounts of food even without taking into consideration the requirements of growth. They must utilize a large proportion of their food merely to keep warm and stay alive. It is clear that, excluding the pigeons and doves in which the adults produce 'crop-milk', the young of birds are perilously dependent upon the day-to-day supply of food that their parents gather for them.

Secondly, the provision of amino-acids essential to animal development must be considered. Most of the comparatively few birds that are largely vegetarian appear compelled to switch at least partially, and sometimes almost wholly, to a more heavily protein diet for their developing young. Protein is the only source of amino-acids essential to animal development. Thus, domestic fowls, which at first sight appear to subsist largely on vegetarian food, do in fact scratch up a considerable amount of animal food, and artificially reared chicks must be given a diet enriched by proteinous foods such as skim-milk, meat or fish meal or soybean. The same occurs with pheasants. Domestic fowls not only require all the amino-acids essential for the development of the rat but one or two others in addition. An interesting observation has been made on the discriminatory power of the insectivorous Rock Warbler when feeding its young. Adult birds ate proffered bread-crumbs but did not take them to the young, which were fed largely on moths.[39]

There is evidence, then, that even essentially vegetarian species, some of which obtain certain protein constituents from grain, find it advantageous to gather animal food, and that such food is probably indispensable for the development of their young. It follows that it is advantageous for the sexual cycle of a species to be so timed that its young will be hatched at the period of maximum appearance of the traditional food on which they are fed. Parents cannot usually go far afield in search of food because the young must be fed constantly and otherwise kept warm lest they perish. Therefore, there must be a great concentration of suitable food in the vicinity of the nest. As yet little precise information appears to be available concerning the food requirements of the young of any species, but it has been shown[141] that the feeding visits of English Robins to nestlings 7 to 14 days old averaged 13·8 per hour. Again, there is evidence[138] that very tiny European Starling nestlings eat more than half their own weight in food per day and that the food requirement rises even more steeply as they grow bigger.

An excellent example of the necessity for the seasonal concentration of considerable quantities of protein food for the young is provided by the Rook.[180] The males of a breeding rookery, possibly of several hundred birds, feed the incubating females, and later the nestlings as well. The food supply within each restricted area must have become sufficiently abundant to be gathered by one bird of each pair before reproduction can successfully take place. We will see later that among certain bower-birds the female only cares for the young, and that her ovulation is delayed until the environment becomes seasonally full of appropriate protein food.

I have elsewhere stressed the view that the ultimate breeding date (as distinct from the initiation of gametogenesis) has no direct connexion with day-length but is dependent on the environmental factors that permit nidification and the rearing of the young.[171] The theory that when days become a certain length birds reach full spermatogenesis, mate, and ovulate is exploded by the fact that the actual breeding date of various species varies from year to year. No matter what time the sexual cycle begins its resurgence, an 'early spring' will advance reproduction and a 'late winter' will retard it. Records kept by the Marsham family in Norfolk for about 200 years indicate that the first young rooks, for example, appear at any time between 23 March and 2 May, with a mean about the middle of April.[158] Some of these annual observations were no doubt inaccurate, but a perusal of natural history journals shows that the actual breeding dates of many species vary within wide limits from year to year according to the vagaries of the environment.

A most precise relationship between ovulation date and the condition of the environment has been evolved by various parasitic cuckoos. No data are yet available concerning gonad states, but it is known that the females watch selected foster-parents building their nests and, after laying has begun, the English Cuckoo deposits her own egg and often takes away an egg of the dupe. The incubation period of Cuckoo eggs is short, and by means of an innate reflex, the blind and naked young cuckoo rids itself of all competition for food supply in the first few hours of its life. Meanwhile, the female Cuckoo has deposited other eggs elsewhere. To achieve successful parasitism she must lay her eggs at the brief critical period when she has watched nest-building finish and egg-laying begin. This requires, of course, a most delicate timing of ovulation in relation to changing environmental events. In Australia, where exist about a dozen parasitic cuckoos, each species has adapted itself to a different group of hosts and thus wasteful, double parasitism is generally avoided. Occasionally, however, two female cuckoos of a single species lay in the same nest. Further, two females of different species may even more rarely lay in the nest of a bird that is traditionally host to both. This gives rise to a most fierce and interesting intra- and more occasionally inter-specific competition for food, since it is inevitable that only one young parasite shall survive. This matter, however, is outside the main subject of the moment. The parasitic cuckoos are used here merely to illustrate the extraordinary precision with which some members of a widespread family have adapted their ovulation to environmental stimuli. Certain non-parasitic cuckoos, on the other hand, build nests as do other birds.

I believe that the final 'timing' of the breeding season of many species each spring (or at an otherwise appropriate time as determined by natural selection) is controlled not by a single factor such as photoperiodicity but by a combination of external stimuli (including especially behavioural interactions) which, varying from species to species, operate through the exteroceptors, the central nervous system, and the anterior pituitary of the female when the environment becomes seasonally appropriate for reproduction. Any theory of avian breeding seasons based primarily on internal rhythm and secondarily on *various* environmental changes has the advantage of possible application to all seasonal species whether they breed in the spring, in the depth of winter (as does the Emperor Penguin), or on the equator where light fluctuations (about 2 minutes difference between 'winter' and 'summer') are almost certainly too small to command serious consideration.

We have above considered various neuro-endocrinal aspects of the avian reproductive cycle in relation to the environment in which the birds have evolved and carry out their reproduction. We can now consider the breeding rhythms, display, and allied activities of bower-birds in relation to the foregoing data.

4

SATIN BOWER-BIRD

Ptilonorhynchus violaceus (*Vieillot*)

THE 'Satin-bird' (Pl. 1) lives in the well-watered coastal forests of eastern Australia in a discontinuous range of about 1,800 miles from north of Cairns (lat. 16° 58′ S.), far above the Tropic of Capricorn, to the cool valleys of southern Victoria (lat. 38° 50′ S.) (Fig. 4). From above the Townsville area to the Bunya Mountains there is a gap of about 600 miles in which the species is rare or absent, although it is possible that when this comparatively dry strip is more carefully investigated the Satin-bird will be found in odd pockets of heavily timbered country here and there. This break in range has been brought about by the post-Pleistocene march of xerophilous vegetation in some places to the very coastline. The isolated northern birds are smaller than those inhabiting the forests from southern Queensland to Victoria and constitute an apparently valid sub-species, *Pt. v. minor*.[191]*

The Satin-bird is commonest in the rain-forests and particularly at their fringes (Pl. 4). In the south it is also abundant in more open kinds of forest, provided a fairly thick ground-cover of bracken and shrubbery is available beneath the taller timber. Except in the heavily populated southern extreme of its range, the species is very successful. If it breeds in a locality, the male can be usually very soon observed in short, swift flight between trees above his bower-territory, or heard calling challengingly as he pauses on some favourite eminence. In the Sydney area the Satin-bird has been known to build its avenue-type bower in outer suburban gardens. It is perhaps the commonest of the Australian bower-birds. During its nomadic post-nuptial flocking phase it wanders afield among bay-side fig-trees, so it is perhaps not surprising that the first known specimen, described in 1816, was one that turned up in France.[276] It was probably carried there by one of the few French vessels that put into 'New Holland' from time to time before and after the Napoleonic

---

* In July 1867 a male Satin-bird with greenish-blue eyes, olive-black beak and feet, and a brilliant splash of gold on each wing was shot on the Brisbane River a few miles from the city. Diggles,[72] who described it, said that its coloration suggested a hybrid between the Satin- and Regent-birds but that the explorer A. C. Gregory had stated that he had earlier seen the new bird on a tributary of the Burdekin River. The golden-winged male was described as *Pt. rawnsleyi* after its collector. The skin has disappeared, and such a bird has not been collected since. Rawnsley's Satin-bird was probably a mutant or 'sport'.

wars. The bird was early known to the colonists, and named by them the 'Satin-bird' long before they realized that it built a bower. By the time Gould arrived in Australia in 1839, its display-habits were partly known and he was shown a bower in the Sydney Museum. This had been presented to the Museum by Gould's brother-in-law, Charles Coxen. Excited by the sight of the bower, Gould 'determined to leave no means

FIG. 4. Approximate distribution of Genus *Ptilonorhynchus*.

untried for ascertaining every particular relating to this peculiar feature of the birds' economy'. He visited the now long-vanished 'cedar brushes' of the Liverpool Range and discovered several display-grounds for himself. 'Although this species', he said later, 'has been long known to ornithologists and to the colonists of New South Wales, its extraordinary habits had never been brought before the scientific world until I had the gratification of publishing an account of them after migration from Australia.'[102]

The Satin-bird is about 12½ inches long. It is strikingly dimorphic in the adult. The plumage of the fully mature male is almost uniformly black except that the exposed edges of its feathers have a curious refractive property which transforms them to a beautiful lilac-blue in the sunlight. This glossed plumage reflects light to a degree that makes photographs suggest a pied bird more like a magpie than a bower-bird (Pl. 1). The eyes, of 'awesome intensity'[115] in moments of emotional excitement, are probably the most striking physical feature in either sex. The iris has been variously described as lilac, dark blue, sapphire, light blue, pale blue-violet and, in the very young bird, brown. The eye colour seems to vary in moments of emotion. The iris has a circular blood-vessel near its inner rim. This gives rise to an unusually rich anastomotic vascularization which flushes the basic blue of the iris to a colour akin to magenta. The beak in both sexes is partly feathered, hence the Greek generic name *Ptilonorhynchus*. The unfeathered parts of the beak in the adult male are dull blue with bright greenish-yellow tips. The legs are dull greenish-yellow. The whole dark colour pattern of the male makes for conspicuousness in the forest. The gleaming plumage and eyes are epigamic[223] in function.

The female, on the other hand, is essentially cryptic in coloration. Her head and general dorsal surface in different lights varies between blue-grey and olive-green; the ventral surface is lighter, with olive, pale green, and yellowish tints. Individual feathers are crescent-marked with brown and the wings and tail are various shades of brown. There is a region of rich brown-flecked yellow, emphasized by a wide border of grey, which is exposed only when the female flies or otherwise lifts her wings. The vanes of the flight and tail feathers are a conspicuous and uniform straw-yellow colour. The beak is horn-coloured and the legs and eyes are very similar to or identical with those of the male.

Mayr and Jennings say that the immature male has more pointed tail feathers and browner wings than the adult female and that the wing coverts have ochre or rufous edges. The above authors believe that the very young male moults into a sub-adult plumage which differs from that of the adult female 'by having smaller light spots on throat and breast and by having a definitive greyish-green collar across the lower throat'. Young males 'average larger than females'.[191] The 'collar' mentioned above has indeed been seen in young captive males about to change plumage and the beak, too, is then closer to the male colour than to the female. A richer coloration in the ventral mottling seems also to be a feature of young males, as well as a slightly more bluish dorsal gloss.

There is a widespread belief that all males remain in immature plumage until they are 7 years old. This was conclusively disproved by Hirst[112-13-14] who successfully bred the Satin-bird in captivity. He found that a young male acquired a few dark feathers in its fourth year (at the end of the autumn moult) and completed the transition during the next moult in the following February. Thus, this particular bird came into adult plumage at the age of 4 years and 2 months. On the other hand, it appears that individual variation exists. A bird in the Chicago Zoo[220] was captive there for 6 years and 3 months before becoming completely dark. This bird was received in Chicago in May when it must have been at least 5 months old. Thus, it was almost seven, and possibly older, before its colour-change became complete. It was noticed, too, that this bird's beak changed from a dusky colour to 'pale green' at the time (February) when the first—and solitary—feather appeared. Among my captive experimental birds (to be mentioned later) ages were unknown, but it was found that all males began to change colour within 3 years. Further, all males received by the New York Zoo[60] showed first signs of colour-transition with their first moult there. Sometimes—in either bush or aviary—only a few, or even one, dark feathers appear and then there may be no further change until the next seasonal moult. It seems generally agreed that once begun, the colour-change is completed within a period of 2 years. A male therefore generally spends at least a year of its life in mottled plumage. As will be shown later, sub-adult males come into full breeding condition before assuming even one fleck of the darker plumage, so it seems certain that plumage-change is not a simple transition dictated by the secretion of gonadal hormones. While displaying in the sunshine, the greenish plumage of the immature male seems to take on a mauve gloss, but this is of course quite different from the flashing spectacle provided by the completely adult birds. In the essay which follows, individuals will be referred to simply as blue, green, or mottled birds for the sake of brevity.

The bower of the Satin-bird is formed of two parallel walls of arched dry twigs (Pl. 1) and beside it, usually at one entrance, is a display-ground on which is haphazardly strewn a sometimes startling variety of coloured decorations. These are, however, chosen with great discrimination. At the end of the display season, late in summer (December to February), the testes metamorphose and the bower and display-ground are allowed to fall into disrepair and the family party—male, female, and one, two, or rarely three young—joins others to form large communal flocks which roam the forests in search of food. Noisy and conspicuous, the flocks may number up to 100 strong. Bushmen and others say that they consist of 'one blue bird and about fifty green ones' but careful scrutiny generally reveals a plurality of blue birds in any flock containing more than 20 or 30 individuals. One observer records nine blue birds in a flock of about fifty individuals.[46] Likewise, several males in colour-change, whose mottled plumage is not readily identified in the tree-tops, are always present in any big gathering. This post-nuptial flocking phase is in no essential way different from that which occurs in European birds such as the Rook or Chaffinch or, for that matter, in numerous other Australian species. After the exhaustion of the testis interstitium and the post-nuptial metamorphosis of the seminiferous tubules in the male, and the seasonal deliverance of the female from domestic duties, the parents lose their territorial attachment and wander afield accompanied by the young. The collapsed testes of the male Satin-bird now measure about $2 \cdot 0 \times 0 \cdot 8$ mm. and weigh some 5 mgm. The seminiferous tubules are 40μ in diameter and contain spermatogonia and a quantity of lipoidal debris. The interstitium consists of massed juvenile Leydig cells that measure about $5 \times 5 \mu$ and contain a nucleus 3μ in diameter. The testis tunic is about 100μ thick. (At the height of the breeding season (October–November) the testes measured about $14 \cdot 0 \times 8 \cdot 5$ mm. and weighed as much as 510 mg. each.) By February the ovary as well as the testes has subsided and only the still partially distended oviduct reveals macroscopically that ovulation has taken place.

The flock wanders. There is still plenty of proteinous insect food in the forest and some of it is eaten; but it is to the wild fig-trees, the 'nutmeg' stands, the *Guoia* groves, or to the thickets of black-fruited inkweed and wild raspberry at the scrub edges that the restless, foraging bands of bower-birds make way. During this nomadic phase the flocks sometimes travel considerable distances and appear in fruit-bearing localities where bowers are never built and breeding does not occur. Sometimes they venture right over the Great Dividing Range, pestering fruit-growers more than 100 miles (e.g. Adelong) from the coastal forests. Thus the species is dispersive rather than migratory. Its precise distribution is still unknown.

In many regions the flocks remain the whole year within a few miles of their bower-territories. From time to time a male will revisit his territory to rebuild the bower and indulge in snatches of out-of-season bower-display with freshly plucked flowers; or perhaps even to paint a little. All this occurs while the testes are still small, with their tubules packed with post-nuptial lipids, and with the newly regenerated interstitium still immature. After brief display, the male will desert his territory and return to the flock as suddenly as he left it.

Meanwhile display of a sort occurs within the feeding flock. From it comes a cacophony of croaking and explosive sounds that are impossible to describe adequately, and which are heard consistently nowhere else. This medley of distinctive noise is perhaps chiefly recognitional in function: while it is going on it is impossible for any individual permanently to be separated, however dense is the forest canopy in which the group is feeding. The conspicuous dark plumage of the adult males probably fulfils a similar function during this phase of the cycle. Brief snatches of arboreal display, very similar to the terrestrial bower-display (to be described later), can be also heard and seen. This flock-display involves sudden jumping, neck-stretching, wing-lifting and other gyrations on high branches. Occasionally there is an outburst of the extraordinary whirring rattle that is a principal feature of the later display at the bower. In these mixed and mobile companies the family parties are broken up and the pair-bonds are begun. It is not yet known whether Satin-birds, like Gannets, Rooks, Ravens,[2] and many other species, pair for life. The male is easily trapped at his bower and four ringed blue birds were proved to build bowers on the same ridges from 1938 to 1945. The shyer females are harder to catch but in two instances it was proved that the same pair of birds occupied bower-territories in successive years.

Sexual periodicity and bower-building

Most Satin-birds return to their territories long before the spring. A male examined at the time of bower-building had testes measuring about $3 \cdot 0 \times 3 \cdot 3$ mm. (Pl. 7). The tubules contained a prolific growth of primary spermatocytes and no lipid material whatever. The new germinal growth had swollen the tubules to a diameter of $80\,\mu$. The interstitium contained cells which were still juvenile in character but most were larger ($9 \times 9\,\mu$) and more strongly lipoidal and positive for cholesterol. The tubule development had stretched the testis tunic which was now only $50\,\mu$ in width. The earliest personal records of the construction of the 'permanent' seasonal bower to which a female is brought, are May, the last month of autumn. Nubling[208, 209] has recorded the erection late in April of a bower that 'lasted until 30th December' but it is not known whether display at it was consistent from the time of its construction. The male may either renovate thoroughly his old structure or, probably more often, build a completely new bower on the same site or close by. In this he often incorporates any twigs that still remain in the old one. When this happens the old twigs are not pulled up but are broken off at platform level and carried, two or three at a time, to the new site. Any display-objects that retain their colours are transferred long before the new structure is complete.

Most published accounts of bower-building refer to much later in the year (September onwards) but such information is a reflection of the habits of naturalists rather than those of the birds, since few observers consistently visit the building area before the spring. The flocks do not disintegrate simultaneously as though obeying some regular external signal such as the advent of a certain specific day-length, or the change-over from decreasing to lengthening days, or the summation of a certain light increment from the solstice. Individual males build their bowers at different times within a single restricted area exposed to, as far as is known, the same environmental fluctuations. It is unusual for the bower to be rebuilt as a permanent structure as early as May. More bowers are built in June. Most birds begin to build in July but many others delay doing so until August and a few may not build until early in September. Table I shows the time of construction of forty-two principal bowers whose erection, as apart from renovation, was observed during 1938–40.

TABLE I

May	June	July	August	September
3	8	18	10	3

Such an irregularity, within a single locality, seems to lead to the conclusion expressed in Chapter 2 and elsewhere[171, 180] that the fundamental factor in the control of the annual sexual resurgence of many species is the internal rhythm of the individual on which is seasonally imposed a complex of environmental (including behavioural) factors and not merely the annual advent of some aspect of photo-periodicity. An outline of the gonad cycle of the typical passerine male has been given in Chapter 2. So far the female cycle of only one species (the Rook) has been studied intensively and in this no period was found that corresponded with the male refractory phase. However, if the reproductive apparatus of *one* sex undergoes a phase of negativity then obviously breeding cannot occur until this is over, irrespective of the condition of the mate. It will be only when the male refractory period has ended that he can be stimulated by environmental factors and begin his seasonal sexual resurgence. He can then display and influence the reproductive cycle of the female.

We do not know what specific external factors operate in the case of the Satin-bird—or any other bird for that matter. The generally accepted theory is that light fluctuation of a kind appropriate to the species starts off the seasonal cycle. The winter solstice near Sydney occurs between 21 and 23 June when the day-length is 9 hours 48 minutes. The day-length on 21 May (about when some bowers are built) is 10 hours 10 minutes. Day-length is decreasing at a rate of 1 minute 10 seconds daily. Most bowers are completed by 21 August when the days are a little over 11 hours in length. If it is claimed that, similar to the Suffolk sheep,[290] Satin-birds begin their sexual resurgence as a response to *decreasing* day-length it would be expected that the whole population would begin to build before the solstice whereas, in fact, many wait until day-length is increasing at a rate of 1 minute 10 seconds per day. If it were claimed that the cycle is set off by *increasing* day-length we must explain why some birds build their bowers before, or about, the winter solstice.

Satin-birds, in common with many other southern species, usually modify their display season and come into line with local species when transferred across the equator to Europe.[15, 184] In March 1949 a group arrived at the London Zoo and within a fortnight the most mature male (a mottled bird) built a bower and began to display at a time when birds in Australia were still flocking and paying only spasmodic visits to their territories. It could be held that the changeover from decreasing to increasing day-length caused this resurgence, but the following factors must be considered. When the birds left Sydney early in February the days were about 13 hours 49 minutes long. The birds crossed the equator when the days were 12 hours 7 minutes long. The day-length in London late in March is about 12 hours 29 minutes. Unless it is assumed that the reproductive cycle is stimulated by an initial decrease, followed by an increase, in day-length, it would seem probable that the sudden liberation of the birds from close confinement exerted influences that culminated in bower-building and display. Other fragmentary data suggest that transported Satin-birds may sometimes retain their Southern Hemisphere display season. It was noticed in the New York Zoo[60] that 'gaps in display' occurred generally from January to June which is, of course, the inactive period in the birds' native forests.

Table II provides data which reveal that the movement to the bower-territory occurs at a time of relatively low temperature and rainfall but, late in June and in July, of increased sunshine. It has been suggested[209] that a seasonal improvement in the fruit supply is responsible for the upsurge in bower-building, but in actual fact there is no change in food supply at the time of the gradual disintegration of the flocks.

That temperature is not altogether unimportant in relation to bower-building is indicated by the behaviour of a mottled Satin-bird in the London Zoo on 19 December (approximately equalling 19 June in the bird's natural habitat), during the especially mild winter of 1949–50. This bird began to show an interest in the provided birch-twigs just before the winter solstice, but instead of building its bower in the open it carried the twigs into the heated shelter and tried to drive them into the concrete floor. The temperature outside was 50° F. The indoor temperature chosen by the bird as suitable for bower-building was about 10 degrees higher. Day-length at London at the winter solstice is only a few minutes over 8 hours. At Sydney, whence the bird came, day-length is 9 hours 48 minutes in mid-winter.

It is impossible not to be impressed by the general similarity between the sexual cycle of the Satin-bird and that of many common European species. In the Rook, Sparrow, Starling, Mallard, and others the post-nuptial refractory period lasts from about May until about September when there occurs an autumnal display which is soon inhibited by the severe conditions brought by the northern winter.[176] The Satin-bird undergoes a refractory phase which apparently lasts for an approximately similar period. This begins in January or February and appears to end

Table II
Recorded Temperature, Rainfall, and Sunshine at Sydney

	Jan.	Feb.	Mar.	Apr.	May	June	July	Aug.	Sept.	Oct.	Nov.	Dec.	Annual means
Mean Max. Temp. (F.) (90)*	78·3	77·8	75·8	71·4	65·9	61·4	60·0	63·1	67·2	71·3	74·3	77·0	70·3
Mean Min. Temp. (F.) (90)*	64·9	65·1	63·0	57·9	52·1	48·2	45·9	47·5	51·3	55·8	59·6	62·9	56·2
Mean Temp. (F.) (90)*	71·7	71·5	69·4	64·7	59·0	54·8	52·9	55·3	59·3	63·5	66·9	69·9	63·3
Mean Rainfall (in.) (90)*	3·53	3·99	4·86	5·39	5·02	4·68	4·44	2·90	2·80	2·78	2·83	2·98	46·20
Mean Sunshine (hrs.) (28)*	229·9	199·9	201·5	183·7	179·5	164·5	193·5	217·5	220·6	232·5	226·7	231·1	2484·5

* Number of years during which records were kept.

about May or June, but it leads in the milder Australian climate to a winter sexual resurgence and bower-building. An analogous sexual cycle may be the peculiarly recurrent rhythm of the Sooty Tern of Ascension Island. This bird, as has been stated, breeds four times every three years in its equable equatorial environment. If the Satin-bird immediately built its nest and reproduced, instead of first building a bower and indulging in prolonged display, it would have a cycle very similar to that of the sea-bird mentioned above.

It is suggested tentatively, therefore, that the most likely cause of seasonal building is the completion of the refractory period within the animal and the subsequent reactivation of the neuro-endocrine machinery by local weather conditions appropriate to display. At the beginning of their sexual season Satin-birds rarely display in the rain. Display is heightened by sunny weather. The whole problem, however, invites the detailed analysis of the behaviour within individual territories over a period of years in relation to weather fluctuations. It is by no means impossible that some obscure factor such as the time of year when the early sun strikes a particular bower-site may be partially influential in determining the period when the visiting owner will remain there and begin his sustained seasonal display. (See Addenda, note 'A', p. 189.)

Whatever is the precise interaction of internal and external factors that sets off the annual sexual resurgence, it would seem that it is the resultant liberation of gonadal, and perhaps other, hormones that stimulates specialized areas of the central nervous system and calls into operation innate behaviour patterns concerned with the acquisition of a mate, the defence of the territory, the construction of a permanent bower, and subsequent prolonged display there.

The male builds his bower in the space of a day or two. He covers its adjacent display-ground with as many blue, greenish-yellow, brown, and grey decorations as he can find. The female is generally not far away. Many males during this early period seem unwilling to display unless the female is beside or within the bower. In the flock the male was shyer than most of the green birds. On his display-ground he shows little fear of anything. When building his bower he pauses at odd moments as the calls of other birds come to him. He frequently hops quickly away from the bower, then flies swiftly up to a vantage perch whence he calls loudly over his territory. He shows no fear of a man watching from a new and scanty 'hide' a few yards away, though he may first hop up and peer through its walls to investigate. Satisfied that the intruder means no harm, he returns to his bower. He shows only passing interest in birds of other species, but if a male of his own kind comes near he flies at it swiftly and chases it through the forest out of sight. The rival always flees.

Spermatogenesis occurs within a few weeks of the building of the bower. In the autumn flock the small testes contained minute, meagrely lipoidal Leydig cells and the tubules held considerable amounts of cholesterol-positive lipoidal material (Pl. 9 a). By the end of August or early in September the testes measure about 12×8 mm. and contain tubules some $140\,\mu$ in width. Some tubules already contain bunched spermatozoa. Leydig cells now appear only in small angular aggregations at points whose several tubules would otherwise meet (Pl. 9 b). These individually measure $12\times 10\,\mu$, with a nucleus about $6\,\mu$ in diameter. The testis tunic is greatly stretched and is only $20\,\mu$ thick. Pl. 9 c shows the testis at the height of productivity in October and Pl. 9 d illustrates the beginning of the post-nuptial metamorphosis in January.

We have seen (Pl. 1) that the principal bower of the older males is formed of two parallel and sturdy walls of twigs wedged into a platform of twigs or coarse grass laid on the ground. The bower is built in the undergrowth, usually in semi-sunlight and often beside a fallen log. Each wall is composed of dozens of thin twigs. The avenue is four or five inches wide. The walls are 10 to 14 inches high. They are generally a little taller than they are long and are from 3 to 4 inches thick. The twigs which form them often curve inwards near the full extremities and form a partial or even complete arch. Adjacent to the bower, and usually in front of one entrance, is an extensive platform of twigs or coarse grass on which is placed the coloured display-things. This display-ground varies in area and may, after several years' reinforcement by a domineering old blue bird, come to measure as much as 3 feet in diameter and several inches thick. Decorations are never found in the bower itself. If a suitably coloured object is placed experimentally in the bower the bird will remove it to the display-ground. If the bird regards it with disfavour he removes it into the undergrowth out of sight.

The immature green males build similar structures from their third year (in captivity) onwards.[115] Some of these are of a smaller and less substantial nature. If a young green bird's bower is built near a territory under the domination of a blue bird the young bird generally retains few of his decorations. Again, a third kind of display-ground often exists within a few hundred yards of the main bower. This is merely a cleared bare patch of earth several inches in area on which are laid a few twigs and odd sprigs of flowers, cicada excuvia, and so on (Pl. 6). A low-walled stage of bower-construction intermediate between the simple twig-decked arena and the full-sized bower is occasionally found. This bower, too, is frequented by green birds of both sexes and occasionally by the blue bird of the area. The establishment of the primitive display-platforms can be experimentally induced merely by placing a few fragments of blue glass under a favourite feeding-tree during the sexual season. The green birds come quickly to the ground, clear a small space, gather a few twigs and display with the glass. This will be soon carried off by a blue bird. It has been suggested[96] that this primitive platform or arena is built and maintained by the female, but this has not been proved. Laparotomy and subsequent ringing of green males and females showed me that although females frequent the platforms, their behaviour at them was passive. The actual display there was carried out by green males or by a neighbouring blue male which visited them from time to time. It must be emphasized, however, that only two proven females have been ringed. Aviary experience suggests that conclusions regarding the sex of green birds should never be drawn on the basis of behaviour alone. For example, in the absence of a female, early in the season, a proven green male in captivity in England from time to time occupied the typical female position at the bower while the blue male displayed. It may be that this behaviour was dictated by the conditions of captivity. At the brilliantly decorated principal bower in the forest, a couple of green birds may appear early in the season but as the sexual season heightens any interloping green male is chased away by the owner. Only the green female is allowed to remain.

The paraphernalia of display

A superficial view of a blue male's bower suggests that the bird has made an indiscriminate collection of rubbish, chiefly blue in coloration. This alleged lack of discrimination has received publicity in all kinds of curious ways (for a classical example see p. iv). The blue objects on a display-ground always show conspicuously and, because of this, some writers have tended to ignore the birds' relish for display-objects of other colours. Actually, colours of five different groups are the basis of the display. These are (1) blue, (2) greenish-yellow, (3) 'pure' yellow, (4) brown, and (5) grey. Very occasionally a single bleached bone (e.g. fragment of wallaby or rabbit skull) may appear, but this is so rare that its presence may be accidental. I have seen bleached objects only twice in many years of field observation and captive males never take them to their bowers. There are one or two references to them in the literature. It is possible that in each rare case the bone was lying there before the bird built the playground. A highly reflective object—a threepenny piece—was seen once. It was discovered much later that the coin had been put there by a friend who, knowing that one species of bower-bird (the Spotted) collects coins, made the present bird a gift directly on its display-ground.

A well-decorated display-ground at the height of the display season presents a brilliant dash of colour on the otherwise sombre forest floor. It may contain up to 60 or 70 blue parrots' feathers, odd bluebells (*Wahlenbergia*), the blue, violet, or purple blossom of wild *Lobelia*,

Solanum, Dampiera, Hardenbergia, Thelymitra, Dianella (and its blue berries), many fragments of blue glass, blue-patterned crockery, rags, rubber, paper, bus-tickets, chocolate-wrappers, and the odd blue invitation-card to some function or other. Chaffer[40] records a still-living bluish centipede at one bower. It is interesting that although the valleys of the Sydney National Park are periodically full of the flowers of the purple-mint (*Prostanthera*) only a few odd blossoms are brought occasionally to the bowers. A display-ground may contain more than 100 small greenish-yellow bells of the creeper *Billardiera scandens*, quantities of yellow straw or wood-shavings, a stray sprig of mimosa (*Acacia* spp.), a fragment of rock-lily orchid (*Dendrobium speciosum*), a bit of yellowing onion 'peel', or a few *Banksia* leaves that have been collected only when they have decomposed to an appropriate shade of green-tinged yellow. Grey objects such as puff-ball fungi, wasp-nests, pieces of string, and, very often, lengths of the sloughed skins of snakes are used. One bower examined had beside it over fifty brown landshells (*Hadra* spp.). Odd brown cicada excuvia may be included.

If the bower is built near human habitation the ratio of domestic flowers (such as cornflower, delphinium, petunia, cineraria, iris, periwinkle, hyacinth, jacaranda) and manufactured articles rises. Such a display-ground may contain, apart from flowers, such diverse articles as a fragment of blue piano castor, a child's small blue mug, hair ribbons, a blue bordered handkerchief, and, especially, blue-bags stolen from domestic laundries.[40, 160, 209] About thirty such blue-bags, some empty and faded almost white from rain, and others almost full of 'blue', were counted on one particular bower situated near an isolated country hotel. Display-grounds that are built in the depths of the rain-forest or 'jungle' usually have a greater ratio of the land-shells which abound there. Flowers of suitable shades are comparatively rare in the rain-forest. Therefore, decorations are often confined to a few shells, feathers, fungi, odd fragments of snake-skin, and perhaps a few 'kangaroo-apple' (*Solanum* spp.) blossoms. At display-grounds that remain undisturbed and on which bowers are rebuilt year after year, the faded accumulation of several seasons remains embedded in the platform. Durable display-objects such as blue glass are mostly kept in full view but when an article begins to lose its colour it is discarded or allowed to become submerged by the fresh straw and decorations that the bird brings to his territory.

Generally only one green female is seen at the blue male's bower once the display season gets under way (Pl. 5). Energetic display begins at about sunrise, but it can occur at any time between dawn and sunset. It is interrupted by intervals during which the birds feed, bathe in nearby streams, preen, or call from branches above the territory. At the bower the female is shy (probably because of the human observer, not 'coy' as popular writers have said), but she too will sometimes hop right up to a 'hide' and peer at its occupant without fear. She often hesitates in the undergrowth, or in low branches above, before coming to the display-ground. The male stands square on his display-ground and performs tentatively, keeping his attention focused on the female, who now usually comes fluttering or hopping cautiously forward. She perches on a nearby log, or stands directly behind or within the bower. The male now displays fully, often but not always directing his postures towards the female. She watches idly, taking absolutely no active part in the ritual. The male picks up a bright blue parrot-feather or brown snail-shell in his beak. He begins a rhythmic whirring noise that resembles the sounds made by a mechanical toy. He arches his tail fanwise, stiffens his wings, and holds his head low with neck extended. His plumage glistens magnificently in the sunlight (Pl. 1) and his blue eyes seem to bulge and are suffused with rose-red. The female starts convulsively from time to time when the light flashes from his plumage as he leaps in the sunshine. She may utter a few odd low guttural notes but she is usually silent. Occasionally she will rearrange disordered bower-twigs while the display is going on. The male's mechanical rhythmic 'whirring' continues, accented by a rapid shutter-like movement of wings and tail. A performance may last for a few seconds or as long as 30 minutes. The male shies violently sideways, hops or runs quickly forwards, and pauses, staring fixedly forwards. He will then suddenly resume the display by snatching up another display-object and hopping stiffly around (Pl. 5). As he holds the feather or shell his actions seem threatening, but the threat-gesture is directed at the object he is holding, not at the watching female. The decoration in his beak does not detract from his vocal efficiency. Some green males display as energetically and elaborately as do the older blue birds.

Often the female hops away, or may be frightened from the bower in the middle of the display. If this occurs early in the season the male stops displaying immediately, and calls to her with a curious twice-repeated creaking note which seems exclusively to be used at this time and situation. The female usually reacts to this call by hopping cautiously back and the display continues. Although many people have watched the bower performance, coition has never been observed. Usually the display ends when the female hops away and then flies off through the forest. Later in the season the male may remain and busy himself with renovating the bower walls. He rearranges the twigs with a curious downward sliding movement of the neck. Presently he departs, and sooner or later reappears a little ahead of the female.

If the female attends by the bower while the male is absent she too often busies herself with rearranging the twigs of the bower walls,

However, captive females, continuously provided with suitable twigs and coloured display-things taken from the bowers of males, made no attempt to build or display. A solitary female, however, caged next to a displaying male, will bring coloured material towards his display-ground and sit passively nearby.

The male spends much of his time on lofty perches above the bower preening his feathers in the sunshine. He indulges in brief snatches of arboreal display. Here too, vocal mimicry of other species occurs. The Satin-bird is far inferior to the Spotted Bower-bird in its mimicry but in captivity it will imitate a postman's whistle sufficiently well to deceive a housewife. It will mimic domestic cats, and one bird is known to have continued to imitate wild birds two years after it was made captive.[205]

From the vantage point above the bower, this mimicry probably reinforces the ringing territorial cry (a long-drawn *ee-ooo* which varies regionally), the bold plumage, and the arboreal displays as a long-range threat to marauding rivals.

Bower orientation

The bowers are generally built with the central passage running approximately north and south.[40, 160-1] Of sixty-six principal bowers examined in the bush during 1938–40 the deviation from 360° was never more than 30°. The rudimentary and temporary sub-bower of young green birds, though usually orientated in the fashion of the adults, showed much greater variation. The problem of bower orientation was attacked experimentally in the aviary. In July a blue bower-builder was transferred to another pen and his platform, bower, and many decorations were lifted bodily and set down on precisely its original bearing in the new environment. The owner flew immediately to the bower, and quickly and deftly rearranged disturbed twigs and display-things. No attempt was made to demolish the bower or to build a new one elsewhere. A week later the bower was reorientated by me from 15° to a new bearing of 310°, i.e. approximately north-west by south-east. The owner again flew down and rearranged the disturbed twigs and ornaments. He seemed unaware that any change had taken place. A few days later, however, the bower was partially reorientated by the addition of a substantial third wall. New sticks were used as well as those taken from an original wall which now became a flimsy appendage. The other wall of the original structure was slightly modified to bring it into alignment with the new one. The flimsy partly-demolished wall now protruded from the altered structure in a most peculiar manner. A week later the bird had incorporated almost all of the material of the unwanted wall into the new one and only a suspicion of the demolished wall remained. The change-over is expressed diagrammatically in Fig. 5.

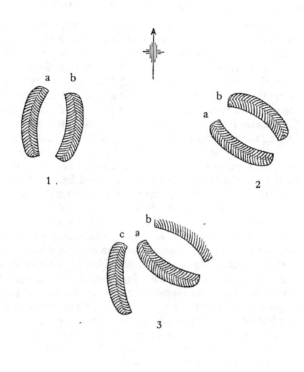

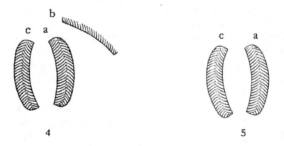

FIG. 5. Experimental reorientation of the bower of the Satin-bird.

1. The original walls, *a, b* on compass bearing of 15°.
2. The same experimentally changed to a bearing of 310°.
3. Wall *a* left intact, wall *b* partially demolished and a new wall *c* built by the bird on a bearing almost the same as the original position of *a*.
4. Wall *b* further dismantled, and *a* straightened to become parallel with new wall *c*.
5. Original wall *b* has been removed. The bower is once more on an approximately north–south bearing.

A fortnight later the bower was experimentally reorientated to an east–west direction. Within forty-eight hours the bird had demolished it completely and rebuilt on a 354° bearing on the same site. The experiment was repeated in the bush the following September when the original alteration involved a swing from 350° to 50°. The wild bird promptly demolished one wall, built another, altered a third, and swung the bower to very nearly its former position. The following week-end this bower, too, was experimentally reorientated due east and west. During the next week it was wrecked and the site deserted altogether.

On the other hand, a blue bird in the London Zoo is recorded to have built its bower east and west.[199] So also did Hirst's young aviary-bred captive male. This bird had to be separated from its father for its own protection in November. It was put in an aviary which was almost wholly built into a sunless artificial cave. Later, when the bird was put into an open aviary, it built a normal north–south structure. Sometimes a bush bower is found pointing east–west. Occasionally one is found with a double wall. The few imperfect bowers built by adult captive castrates were more accurately orientated than those of many intact young birds.

The male probably orientates his bower by reference to the apparent movement of the sun. He becomes more particular, and more expert, as he gains experience and maturity. The utility of north–south orientation may be that very early each morning, when energetic display begins, the male can keep the motionless female in view without staring straight into the rising sun. Likewise she can watch his flashing display without discomfort. The situation is expressed diagrammatically in Fig. 6.

Hirst[115] made a significant observation concerning bower orientation. A blue bird which habitually built its bower pointing north and south suddenly demolished the bower, and then rebuilt it parallel to the cage wire (which was north-west by south-east) when a new female was put in the adjacent aviary. Shortly afterwards he rebuilt again: this time on a north-east by south-west bearing, thus placing the bower exactly at right angles to the wire and the female beyond it. He could now face both bower and female while displaying.

In bowers built in the depths of Queensland rain-forests there may be too many saplings around to allow the construction of an extensive display-ground at one end of the bower. In the dim scrub a bower may have its display-ground beside one wall, or occasionally almost completely surrounding it. Such jungle bowers are generally built north and south. It is assumed that the birds orientate by reference to the bright bars of light which penetrate the gloomy rain-forest.

Functions of the bower and display

We have seen that the bower and display-ground are the focal point of the male's territory. His display there keeps the female within his orbit and keeps his rivals away from her. If the male of the pair is experimentally removed from the bower, the female will remain in the vicinity. She will often visit the bower and perhaps bring odd decorations to the display-ground. Presently she and the bower and territory are annexed by the most aggressive green male in the area. This sequence has been arranged twice. On the first occasion (29 September) the female was

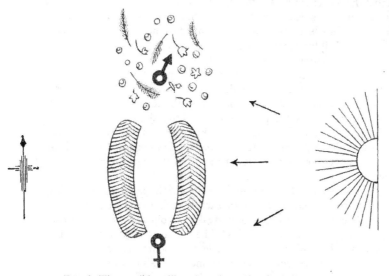

Fig. 6. The possible utility of north–south orientation.

The female Satin-bird (♀) usually stands behind or within the bower-avenue and stares intently at the male (♂) which postures on the display-ground with his coat flashing in the sunshine. He keeps the female in view during the display. In the early morning, when display begins, each is able to look at the other without being obliged to stare into the sun.

mated to a blue male and after his removal she and the territory were both taken over by a green male. This second male was collected on 19 October. Pl. 7 shows the remarkable similarity in gonad condition between the mature blue owner and the green usurper. It is unfortunate that we do not know the precise reproductive state of the second male on the day it took over the bower, but when collected its testes contained large bunches of sperms even though the bird was wholly green and 'immature'. A fortnight after the removal of the second male, the female and bower had attracted a third male. During the intervening period the bower was kept repaired and tidy, but was not painted. The third male was also green. He repainted and redecorated the playground and it was almost certainly with him that the female eventually bred.

In a second experiment a blue owner was removed from his bower and another green male took over the absent bird's mate and possessions within a week. These apparently stayed together, and the female duly ovulated. It was of outstanding interest that in each case the substitute bird was found to be about equal in overall testis size, and in spermatogenetic condition, to the bower-owner that was supplanted. It would seem that the bower is a means to a speedy remating and synchronization of the sexual processes of a new pair if either original bird is killed.

When the female of the pair is removed at a pre-ovulation stage (Pl. 8) (e.g. 27 July) the male will cease to display and will leave the playground partly unattended until, usually within a few days, he finds another female and brings her to the bower. His previous courtship behaviour is now repeated. On the single occasion on which the second female was removed (17 August) the bower was found wrecked a week later. It and the display-things were removed to a site some hundreds of yards away. Apparently the repeated disruption of his arrangements caused the male to desert a territory that he had occupied for at least three seasons. A third female was attracted to the new bower within a fortnight.

During the period of pre-nuptial gonad development neighbouring blue males are continually aware of each other's movements and raid each other's bowers and wreck and steal the coloured display-objects whenever they are left unguarded.[159] The actual wrecking process has been photographed (Pl. 1). When a male approaches his own bower he always flies first to the ground some distance away. Then he hops along a special concealed route through the undergrowth. He never varies his method of approach. A wrecking rival will also hop in stealthily, rather than fly boldly through the open timber. But the intruder comes by the most direct route from his position at the time he observes that the bower is undefended. At the bower the marauder works swiftly and silently and tears down beakfuls of the walls and strews them about in disorder. A wrecker rarely completes his task before he is disturbed by the swift swish of wings of the owner. Usually he snatches up a beakful of blue feathers or glass as he flees. He never stays to fight.

Observation of this aspect of territorial rivalry made it possible to check the extent of the wanderings of individual males during the sexual season. Bottles of an approximately royal-blue colour were smashed, and the fragments numbered with a diamond-pencil. They were then placed in recorded groups in conspicuous positions near various bowers over an area of about 50 square miles. Glass is uniquely suitable in that it can be permanently marked and, unlike most decorations, remains unchanged in all weathers and never loses its attraction. Thirty-four per cent. of the fragments appeared on bowers within 7 days and only 21 per cent. were absent from known bowers one month later. Weekly inspections during the next 2 years made it possible to record the migrations of glass from bower to bower. In two seasons not a single blue fragment travelled more than 1,000 yards. It was found that blue bird A will steal blue objects from B if his own bower is within about half a mile of that of B. Blue bird A, however, will not move across B's territory in order to steal from C. Again, if a bird is in possession of a comparatively isolated territory he shows no disposition to venture afield seeking combat. Most of the bower-wrecking and theft takes place when two aggressive blues are forced by food or decoration availability factors to occupy adjacent territories. Two such areas of intense competition occurred within half a mile of each other on forest spurs near the Bird Cabin in the Sydney National Park. Both places are popular luncheon stops for motorists and these provide a week-end abundance of blue objects (cigarette packets, chocolate wrappings, &c.) and food fragments. Thus, colour and food availability allowed four bowers to be built within one abnormally restricted area of about half a mile. Near the Cabin one blue bird occupied a ridge and the second an escarpment only 70 yards below. Both bowers were well decorated, and raiding and counter-raiding were continuous. About half a mile upstream a third blue bird had a colourful bower, but a nearby fourth bird appeared to be less possessive and combative, and exhibited a comparatively meagre display. The actual display performance of the fourth bird was less sustained than is usual, and it was thought that he may have been aged or sick. Certainly he did not effectively compete with his neighbour in the retention of display-objects.

Near the Cabin each blue bird dominated territory only in the immediate vicinity of his respective bower. Neither appeared to venture far away. They would meet on common ground at the Cabin whence they came to eat food and scraps placed on a stump. But while one bird was feeding, the other waited and flew to the food-stump only when it was vacated. If, however, either approached the other's display-ground the owner would swoop swiftly towards the trespasser, which always fled. Despite their individual ferocity, the marked glass and other decorations made repeated journeys up and down the ridge to rival bower and back again. Here a remarkably even balance was maintained, whereas at the upstream bowers, as has been mentioned, there was an almost exclusively one-way traffic in stolen display-things. The impression is gained that in the Satin-bird the utility of the territory is not primarily to ensure a fairly even distribution of the population over the available food supply, but rather to establish a restricted area to which a female can be attracted and guarded until fertilization time. Bourke and Austin[28] found two nests built within an area of only 50 yards in the upland rain-forests of northern Queensland where the species is not notably common. If there was a fierce intra-specific struggle for food it would seem that a greater separation would be desirable.

It has been seen, then, that once display activity is well established the male Satin-bird becomes rigidly and uncompromisingly territorial. Hirst[115] says that at the height of the display season a blue male attacked and badly injured its own green male offspring which, in captivity, could not escape. Associated with this physical rivalry is a driving urge for display-things which all male birds, blue or green, try to obtain for their terrestrial (as opposed to arboreal) display. The old male, exercising complete domination over all green birds in his area, is able to appropriate any suitable decorations which the green birds collect for themselves. He has, therefore, indirect and involuntary collaborators in the decoration of his display-ground, for he will allow the green birds to retain nothing that he personally finds attractive. In the Macpherson Range rain-forest, green Satin-birds of undetermined sex were particularly troublesome at the O'Reilly homestead because of their seasonal thefts of blue delphiniums and petunias. The green birds were bolder than the blue birds in their raiding. They wrestled with 24-inch 'spikes' of delphiniums within a few yards of the aggrieved, but half-amused, gardeners. Yet I never found a comprehensive colour display on the green birds' bowers a few hundred yards away. The blue flowers always appeared eventually on the bowers of the senior blue birds farther off in the rain-forest—one at a distance of about half a mile. Such was the dominance of the blue birds that they were able to carry off any decoration of value immediately it was placed by a green bird on his display-ground.

This matter, too, was put on an experimental basis. In the spring of 1940, 100 fragments of blue bottles were scratched with a diamond-pencil and distributed overnight among eighteen runways or sub-bowers along the Carrington Drive of the Sydney National Park. Before noon the following day, 76 per cent. of the fragments had appeared on the display-grounds of neighbouring blue males. In two instances actual transference was observed. On the first occasion, at sunrise, a green bird began an excited display when it discovered the unexpected treasure on its platform. It was immediately joined by another green bird and then, within a minute, their noisy reaction attracted a blue bird which came flying low through the *Acacia* grove. It hopped the final few yards to the runway, seized a fragment of glass and displayed briefly while the green birds remained watchfully silent. Then the blue bird flew swiftly, still close to the ground, up the slope to his own bower. He returned again and again until all the blue glass had been transferred. At no time did the green birds attempt to defend their property. At the second bower from which transference was observed, three green birds, two of them actually displaying, were in possession when a blue bird arrived. Again the intruder displayed before making off with a fragment of glass. On investigation it was found that some of the scratched pieces had been removed before my arrival.

The function of colour-display

Various authors[162, 199, 209] have suggested that the preference of the male for blue and greenish-yellow display-objects may be connected in some way with the blue eyes and greenish plumage of the female. Crude experiments to check the colour preference of the Satin-bird have been carried out spasmodically for the past 50 years in both aviary[22] and in the bush.[36, 75, 130, 161] It has been claimed by some that, far from merely ignoring red, the blue bird shows anger when this colour is placed among the others on his bower.

Experiments carried out by Morrison-Scott[199] in the London Zoo showed that of 340 shades (including a series of 14 ranging through greys from white to black), the preference of a blue male 'fell into two groups centred around cornflower-blue and lemon-yellow. No preference was shown for the yellow group as opposed to the blue group.' The same author carried out tests which seem to show conclusively that the brightness of objects is not a significant factor in the birds' selection. He checked also the birds' alleged aversion to red. It was found that whilst the Zoo bird refused to tolerate red on its playground, and refrained from gathering red objects, it did not exhibit the intense objection ascribed to it by naturalists. The Zoo bird removed red test pieces and stuffed them out of sight under the twigs of the platform but it did exactly the same with dark greens and grey in later tests. In crude tests carried out by me in the bush several years earlier the blue male simply picked up the red objects from its display-ground, flew away, and dropped them out of sight in the bush.

In 1939–40, experiments were carried out in Sydney with the object of determining what percentage of red and green the Satin-bird would accept in association with its favourite colours. Cards measuring $3 \times \frac{3}{4}$ inches were painted in various proportions of red-blue, green-blue, red-lemon, green-lemon, red-grey, green-grey, and so on, and were offered to captive birds of both colours and sexes. The tests were held at the height of the display season and were spread over several weeks in order to eliminate the fatigue factor. Whenever the cards were put in the aviaries the males (though they had already blue glass, blue-bags, flowers, and onion and pineapple skins decomposed to a suitable lemon-yellow stage) swooped down with immediate interest. After only a few seconds' hesitation they selected the cards they wanted and carried them to the playground. The females often turned the cards over, or carried them a short distance before dropping them. They never showed sustained interest. This was also the later experience of Hirst,[114] to whom I lent the cards in order to undertake an experiment that will be described later. The Sydney aviary tests accidentally brought to light a fact that had been apparent at bush bowers but which, curiously, had never been commented on—that, although the males are elastic in their choice of blue shades, they are most conservative in their selection of articles coloured greenish-yellow, grey, and brown. Although the colour of the greenish-yellow cards very closely approached that of *Billardiera*, the grey that of puff-ball fungi, and the brown that of fresh *Hadra* shells, the birds treated them as though they were red and green and ignored them.

In the percentage tests, the male birds showed an almost mathematical degree of discrimination in their successive preferences. The males almost invariably snatched first the cards with the greatest percentage of blue (75%), then the card with the next biggest proportion (50%) and then the card with 25 per cent. of blue. They would take to the playground a red card bearing a thin end-strip of blue (10%) even when the playground was already almost covered with that colour. This argued eloquently against an *aversion* to red. No obvious differences in relative preferences were noticed between the blue and the younger green males.

Nubling[209] was probably the first person to realize clearly that the collection and display of grey and brown, as well as blue and greenish-yellow, objects may have its genesis in the resemblance these colours bear to those of the female eyes and plumage. Nubling declared that the collection of brown shells and cicada cases, and the dry olive-green leaves and the umber fungi, can be explained by the presence of identical sombre shades in the plumage of the female. He supplied detailed tables, the result of many years of observation, which seem to show conclusively that there is a far closer correlation between male decorations and the less spectacular parts of the female's plumage than anybody had previously suspected.

At the same time it must not be imagined that the males use exclusively decorations whose colours match exactly those of the female. Such is not the case. It is easily demonstrated that many shades of blue taken to the bower (both experimentally and naturally) do not precisely match the female iris. The widest limits of colour variations of the iris under emotional stress do not compare with the differences exhibited by blue display-things which may often be found on a single playground. The selected shades of yellow do not always correspond precisely with those of the plumage. Again, the more sombre display-things such as snake-skins, fungi, leaves and the rest by no means always exactly match the obscurer tones of the female plumage. To prove this, a freshly killed female was carried from bower to bower so that a precise comparison between display-things and the bird, including especially the soft parts, could be made. The blue decorations examined included a wide range of shades of blue such as those of pale *Solanum* and dark *Hardenbergia* blossoms, as well as many shades of blue in tinfoil, broken glass, crockery, and synthetic ware. *None* of the shades on the bower matched exactly the female iris. A manufactured object, a bottle fragment, approached it closest. Later, the eyes of a living female were compared with the objects on a bower. The parts of the iris which in the living bird flush reddish were still less like any of the display-things. The sprigs of *Acacia* (the only 'pure' yellow found on these bowers) also matched imperfectly the greenish-yellow and clear yellow tones of the female. Of the two species of shells found on the bowers, one matched the brown of the flight feathers to a very close degree; the second rather less so. One greenish-tinged shell was similar to a shade of the wing-coverts.

Nevertheless, a close resemblance between the colours of female and display-things undoubtedly exists. This being so it might well be asked why there are not more published records of moulted yellow feathers of the green birds being found at bowers. The reason is that the moult in both sexes occurs during the beginning of the post-nuptial gonad metamorphosis which is the time of year when the bowers are deserted and later only visited spasmodically. When birds are confined in aviaries and cannot therefore move nomadically from their bower-sites, moulted feathers are used regularly. At Bart's, in the autumn (early in September), my birds use such feathers. Further, a young male that had recently changed plumage used its own flight feathers which are predominantly blue-black, but which have still a small portion of the yellow neutral plumage colour near the base of each plume.

When acceptable blue and greenish-yellow objects are experimentally placed near the bower, the blue article is usually taken first. At the same time careful analysis shows that many bowers have a greater proportion of greenish-yellow decorations than blue. Two bowers in the upland rain-forest of North Queensland were decorated exclusively with 'a pale green fruit'.[28] It was not reported whether blue and greenish-yellow objects were available in that particular forest nor, unfortunately, how closely the pale-green fruit matched certain greenish tones in the female.*
On the other hand, local abundance of certain of the less colourful (as distinct from blue and lemon-yellow) display-things does not result in the display-ground being swamped with them. In October stupendous numbers of cicadas emerge from the ground and almost every tree-trunk

---

* Or those of the male. Latest experiments (see Addenda, note 'B', p. 190) suggest that this indeed is the case.

is festooned with their excuvia. Only a few of these are taken to bowers. Individual, or local racial, preference also may operate to some degree. Birds in some areas use extensively the purplish flowers of *Hardenbergia*. Those of others appear rarely to do so even when it is one of the commonest flowers of the region.

It appears too, that apart from mere colour, form probably plays a part in the male's choice. A traditional preference for the object itself, as well as its colour, seems to operate. Thus, snail-shells, cicada excuvia, fungi, and lengths of sloughed snake-skin are brought to display-grounds whereas sticks, dry leaves, and stones, which sometimes more closely match the female colours, are not.

There has been speculation[162, 199, 209] in the past whether the Satinbird's predilection for blue and greenish-yellow is the result of free choice or an automatic selection due to the anatomical composition of its retinae. As long ago as 1866 it was shown[250] that cones of the avian retina contain coloured globules that are filters through which light rays must pass before stimuli are received by the optic nerves. Hess[109] reported that fowls, placed in proximity to a strong spectrum illuminating grains of rice, ate only those which were red, yellow, or green. The grains coloured green-blue, blue, or violet were neglected. It was suggested that the fowl's cones contain only a restricted range of coloured globules and the birds are consequently colour-blind to certain wave-lengths. These results led another writer[78] to deny the use of blue decorations by Satinbirds. Again, later work[214, 235] showed that although the photoreceptors of cocks contain oil globules of several colours, blue is apparently rare or absent. So, although as early as 1907[289] it was shown that the retinae of at least one passerine (a finch) possessed blue droplets, it was for a time fairly generally believed that almost all diurnal birds are blue-blind. Although in the Satin-bird this question of choice cannot be settled without anatomical investigation of the retina, there is ample evidence to support a working hypothesis that the males have a free-choice colour vision that is wide in range and of high acuity. Their choice of a broad range of the shades of blue, their conservatism in regard to the shades of other chosen colours, and their complete rejection of still others seem to point to this conclusion. It could, of course, be suggested that the species, along with its evolution of the female eye and plumage colours, has developed a retinal apparatus that ensures the exclusive choice of the colours which match them. Such an hypothesis, however, seems unnecessarily complicated.

Morrison-Scott suggested that the male may like to play with greenish-yellow and blue objects 'as kinds of *Ersatzobjekte*' for the female in her absence, and that they might exercise an excitatory effect on the male by virtue of their resemblance to the female colours. The second part of this notion was an early personal[162] conclusion. It has since been modified and extended as will be shown on pp. 61 and 190.

Bower-painting

The younger males all seem to enjoy the collection and display of coloured objects; but the same cannot be said of yet another remarkable seasonal activity—the painting or plastering of the inner walls of the bower. This habit appears to vary among individuals irrespective of colour and maturity. It apparently remained unobserved until 1920 when Nubling[207-8] saw National Park birds applying some sort of masticated material to the inner twigs of their bowers. He decided, on further observation, that fruit-pulp was the material used. Gilbert,[95] too, observed Satin-birds painting their bowers with the pulped fruit of the blue-berry lily (*Dianella*), blue-berry ash (*Eloeocarpus*), native plum (*Sideroxylum*), geebung (*Persoonia*), and other unidentified plants. Gilbert considers that apart from their original decorative effect, some pulps induce the growth of a culture of soot-like fungus which clings to the twigs. Meanwhile, in 1922, a blue male in New York Zoo was noticed following a keeper about the aviary, nibbling particles of soft wood from the sieve that the man was using. The bird then chewed the wood until it was thoroughly mixed with saliva. It finally smeared the resultant paste on the inner twigs of the walls of the bower. The result when dry was a thick greyish coat of crumbly consistency. 'A piece of dry rotten wood was ... placed in the cage and an orgy of plastering followed.'[60]

In 1930 Gannon[85] described how a blue male used a small wad of bark as a tool to assist in painting his bower with a suspension of chewed charcoal in saliva. Many people have since observed this remarkable phenomenon. Some—but apparently not all—males collect fragments of fibrous bark, manipulate them with the beak and so manufacture a small oval pellet which measures about 10×6 mm. and 4 mm. thick. This is not a brush. It seems to be a kind of combination sponge, wedge, and stopper which is held almost wholly within the beak somewhat towards the distal extremity. The pellet keeps the beak slightly open as the bird jabs at, and paints, the individual twigs with the sides of its beak. It acts also as a sponge to help retain the liquid while this operation takes place. As the bird applies its beak to the twigs, the charcoal-saturated saliva oozes from between the mandibles and covers the twigs with a thick, sticky, uneven jet-black plaster to a thickness of 2 or 3 mm. The plaster soon dries to a gritty charcoal powder which will rub off on the finger, and which is washed away with every considerable fall of rain. The paint is replaced daily during the height of the sexual season. Logs charred by bush-fires are plentiful at the edge of the rain-forest, and so in most localities the birds have a never-failing supply of painting material.

I, and several others, have stimulated impromptu paintings merely by placing fragments of charcoal on the bower-platform of a known painter at the height of the season. In these planned performances one can actually hear the charcoal being ground up between the mandibles. Under the microscope, solid grains of charcoal as large as 0·5 mm. can be measured.

The wads of bark, still saturated with charcoal and saliva, are often left discarded on the floor of the avenue. Decorations, on the other hand, are never placed within the bower, nor is a fallen leaf allowed to remain there an instant after the owner arrives.

Materials other than fruit-pulp and charcoal are sometimes used. A substance which had the appearance of finely divided wood was found on a bower by Chaffer[40] on Camberwarra Mountain. In the London Zoo Morrison-Scott[199] observed what may have been a form of paint. 'The bower . . . had the inner wall twigs coated with a blackish mud for a distance of about 8 cm. and the bottom of the "painted" section was about 16 cm. from the floor of the bower.'

In addition to wood-paste, fruit-pulp, charcoal-paste, and possibly mud (in captivity), stolen washing-blue is sometimes used. Twice, at Shellharbour and at Minamurra,[160] I found bowers with their inner walls brightly stained. Since then there have been other accounts of painting with washing-blue. There is also a report of damp blue paper being used to stain a Victorian bower as early in the season as July.[63] A newspaper report[3] of a Queensland painting with washing blue may also be mentioned in passing. The newspaper contributor put some blue near the bower and the blue owner duly used it on the bower walls. The bird employed the blue-bag for several days. After this was exhausted, it reverted to charcoal. The newspaper contributor's bird is said to have looked at the proffered blue with a 'thoughtful expression', and was 'roused to fury' by a piece of red material placed on the ground. There is, however, no reason to doubt the purely descriptive part of his story.

Young green male birds often paint their bowers with fruit-pulp, and some do so with charcoal. One bower, from which the blue male was removed for experimental reasons, was taken over by a green male which appropriated also the blue bird's mate, his decorations, and his painting duties as well. At the same bower it was noted that although the female visited the bower during her brief solitary period she did not paint. In the New York Zoo it was apparently the male alone which painted.

In view of the variety of materials used by both blue and green birds under natural and captive conditions, it is surprising to find that all birds do not seem to paint. Some bush bowers remain unpainted at the height of the season. My captive birds rarely used the materials provided. Hirst reported this too.

There has been a great deal of speculation concerning the origin and function of the plastering habit. One original observer ascribes to the bird powers of 'some kind of deductive reasoning' since it has 'experimented with the various materials offered by Nature' and because, he says, 'on certain occasions intelligence aided the bird in solving problems'.[208] Another is inclined to ascribe a Lamarckian mode of origin.[96] He believes that the bird may have accidentally wiped its beak on the twigs of the wall after playing with decorative fruits and, 'the result being more or less pleasing to this somewhat intelligent and discriminating species, the habit was persisted in and ultimately transmitted to, or acquired by its offspring'. The same writer appears to think, however, that the habit may, once the original bird accidentally plastered the twigs, more probably be passed from father to son, generation to generation, apparently more or less in the manner in which Man transmitted primitive art. It has always been denied by naturalists that the phenomenon is inherent and unconditioned.

I believe that bower-plastering is essentially an extension of the courtship-feeding phenomenon that is so widespread in birds. This question will be considered more fully in *Discussion* at the end of this chapter. Courtship-feeding by the male was observed in the aviary, and an activity that probably comes under the same heading has been described in the Sydney Zoo. Conversely, there is an account of green birds feeding blue and mottled males along with arboreal display by the recipient.[51] Something which may have been courtship feeding has been described by Chaffer[42] in the wild Spotted Bower-bird when, in front of the female at the bower, a male 'occasionally went through a kind of vomiting action ... with open bill and shaking head he seemed to be endeavouring to rid himself of something unpleasant'.

Experimental modification of the sexual cycle

We have seen (Pls. 7, 8, 9) that bower-building and associated display activities run hand in hand with seasonal development of the gonads. Experimental investigation of the sexual cycle of male birds that were known regularly to display threw further light on this important matter. In May 1940 two males which had previously built bowers were castrated at a period of testis regression. They thereafter did not build bowers nor display when a control bird did so during the following winter and spring. It was proposed, after a year's recovery period, to inject the experimental birds with testosterone to discover whether display and bower-building could be re-established; but the birds escaped early in 1941, and the war put an end to work for the time being. During army leave in October 1944 another bower-building mature male was castrated at the height of the display season and of spermatogenesis. Its bower, and that of the control bird, were experimentally destroyed. The control male rebuilt immediately, but the castrate failed to do so. The following spring, however, the castrate built an imperfect bower, gathered coloured objects that I provided, and displayed vigorously to females in an adjacent aviary. Transferred to a second aviary from which it could not see the females from the ground, it built another abnormal bower, and spent some time each morning displaying from a perch within view of the females. The display then waned and had almost discontinued by October 1945 at a period when control and wild bush birds were at the height of display. The experimental bird's interest in bower, display-objects, and neighbouring females seemed to have declined completely by November in the season following gonadectomy. Its vocal mimicry of other species declined also.

On 13 December, 2·5 mg. of testosterone propionate was administered intramuscularly and 6 days later an additional 4 mg. was given. The following day the bird was liberated in its aviary and it displayed vigorously to the females next door. A new bower was built on the former site; but it, too, was imperfect, having only one wall which ran parallel with the wall of the aviary. By mid-January the castrate had lost interest in its bower and in the females, but a further injection of testosterone caused a renewal of attention at a period when the interest of wild birds in their display-grounds was declining. Two subsequent injections, on 30 January and 11 February 1946, were followed by minor display. After cessation of treatment the bower was allowed to fall into disrepair. In April the bird was dissected and found to contain no trace of testicular tissue.

The nesting season

Neighbouring species (Yellow-tailed Thornbill, Rock Warbler, New Holland Honey-eater) which begin their seasonal sexual development at about the same time as the Satin-bird have produced young by August. The displaying male Satin-bird is now approaching full spermatogenesis. The cycle of the attendant female lags behind that of the male, although ovarian development has proceeded a considerable distance. Early in September further numerous neighbouring open-forest and heath-dwelling species are mating. The Satin-birds still make no attempt to do so. Their stomachs still contain mostly nutmegs, wild grapes, figs, and ink-berries. The males still display energetically in front of each highly decorated bower. The female remains in attendance, but no attempt is made to copulate. The physical attention of the male still seems to be directed solely towards the ornaments and the bower. The female watches him intently, but that is all. Then, towards the end of September, while the male display continues as before, the female can now be seen gathering long thin twigs of the kind that she uses for the framework of her nest. Birds of many species can build a most elaborately woven nest in the course of a few days but the female Satin-bird spends many days apparently aimlessly gathering and discarding nest material.[94, 96] It has been reported that during this period the female will drop nesting-twigs in mid-air, or may place them on branches that provide no possible support for a nest.[225] Ultimately she builds a simple, shallow, saucer-shaped, leaf-lined structure which she will possibly desert.[83] She may partially complete or even complete two or three nests before ovulation.[123]

Sometimes in October, but generally in November or December, the female leaves the bower and lays her customary two, sometimes one, or, very rarely, three eggs in a nest that she has built usually a few hundred yards away from the bower. The nest is lined with flat eucalyptus or sally wattle (*Acacia*) leaves, and when these dry, the cream or buff, dark-spotted, blotched or streaked, cryptically coloured eggs merge harmoniously with their leafy background. The nest may be built at any height between 4[51] and 115 feet[225] from the ground. It is usually placed in topmost branches, or in a bunch of mistletoe, about 40 feet aloft in the open forest (Pl. 6). Sometimes in Queensland the Satin-bird will build in a tree that is more or less isolated and conspicuous, even though her bower is concealed in a dense growth of rain-forest close at hand. Occasionally the nest is built in thick jungle. The male takes no part in nidification. He does not assist in the 18–19[96] to 23-day[83] incubation, nor does he feed the female on the nest. Protected by her cryptic plumage she alone carries out all the domestic duties. Fleay,[83] who bred the species in captivity, says that when he approached the nest the female showed great alarm, including distraction display ('broken-wing' mechanism and mimicry of other species), whilst the male exhibited no concern whatever.

After the female leaves the bower to build her nest, the male continues his display with considerable fervour. There is no evidence that a second female comes to the bower. In view of the dominance that the blue male holds over the immature green males at the sub-bowers it seems possible that he may there fertilize a second female and so polygamy may occur. This, however, has not been proved.

Neither is it known what events cause sudden transference of the physical attention of the male from his display-things to the more or less similarly coloured female which has been waiting alongside for so many weeks. The blue male will display to, and attempt copulation with, a mounted female skin placed at the edge of the display-ground, but so far this experiment has been carried out only under aviary conditions in October. On another occasion the skin was attacked and sent sprawling. Copulation was not attempted when the skin was placed 20 feet away from, and out of sight of, the bower. Further experiments, with the female skin in various postures, are being conducted.

It will be recalled that some Satin-birds begin to display in May or June—yet ovulation rarely occurs before the end of September, and generally one or two months later. What is the cause, and utility, of such prolonged delay? The fact that the male's testes have contained spermatozoa for so many weeks makes it probable that it is from the female that the final sign-stimulus to consummation comes, and that this activity, whatever precisely it is, has been in turn initiated by environmental events. During her prolonged attendance at the bower, her neuro-endocrinal apparatus is no doubt stimulated by male display, and so sexual synchronization of the pair comes about. But her dilatory nest-building activity seems to make it clear that her central nervous system and internal reproductive apparatus still lacked some essential end-stimulus or stimuli that would allow her to mate. Part of this complex may be the seasonal advent of a special food used to feed the young.[162]

The utility of the delayed ovulation may be explained as follows: it is advantageous for the adult male to annex a mate early in the season and to bring her to his exclusive territory where he is seemingly certain to retain her. The display keeps her within his orbit and absorbs his energies until she permits coition. However, although adult Satin-birds are largely frugivorous, their young, like those of other essentially vegetarian species, cannot come to full maturity without the provision of essential amino-acids. This was proved by Hirst. For four successive years his aviary-bred chicks died within a fortnight of hatching because their mother could not provide enough suitable protein food.

The specialized feeding habits of the Satin-bird make it necessary for a considerable concentration of flying protein food to be available when the young are in the nest. Captive adults will eat avidly many sorts of insects. But should proffered land-crustaceans (for example) crawl under leaves in the aviary, the Satin-bird lacks the capacity to seek them out. Yet flying insects, which the birds can keep under constant observation, are pursued with the greatest determination. In the bush, Satin-birds eat flying food such as termites, beetles, moths, and cicadas.* Satin-birds which build beside tidal rivers forage in nearby swamps when the mangroves are in flower. I suspect that it is the myriads of honey-sucking insects that is the attraction. Satin-birds capture also terrestrial insects; but they do not scratch aside fallen leaves, nor search under loose bark, nor scatter loose earth, nor engage in the other manifold food-searching

---

* There is a record of a Satin-bird catching a 5-inch lizard.[269] Captives prefer minced horse-flesh to fruit if deprived of proteins for a few days. Despite their bifid, brushed tongue they have not been seen to drink nectar.

activities common to many other passerine birds. Thus a tremendous amount of protein food is inaccessible to them.

In September and October dense swarms of termites (*Coptotermes* spp.) suddenly take to the air, as do also Bogong moths (*Agrotus infusa*) in countless numbers. Towards the end of October and in November cicadas emerge from the ground, usually in numberless hordes. In December, 'Christmas beetles' (*Anoplognathus* spp.) appear and attack the eucalyptus foliage in stupendous numbers.[201] All these insects the Satin-bird eats.

On the other hand, it must be emphasized that no evidence, experimental or otherwise, is available to suggest that it is the actual ingestion of the seasonal flying insect food that acts as a direct stimulus to mating. Indeed, it must be recalled that Hirst's captive pair mated and reproduced several years in succession at the normal times while prevented from catching the natural food outside the aviary, although, no doubt, some insect food strayed through the wire and was eaten. It does seem certain, however, that whatever factors stimulate mating and fertilization, these events have been naturally selected and so lead to the launching of the young into an environment that is teeming with insect life of the kind on which they are traditionally fed. As the female receives no help from the male in the feeding of herself or the young, it is essential that there should be a considerable concentration of appropriate food within easy access of the periodically unprotected nest. It will be recalled, too, that the nest is a shallow saucer of twigs and leaves that is quite unsuitable for heat conservation. Thus the eggs must be laid, and the naked young hatched, at times when the external temperature is sufficiently high to permit absence for food gathering. It is possible, therefore, that changes in temperature influence the female to nest. However, the mean minimum temperature in October is only 4·5° F. higher than that of September, and that of November only 3·8° F. higher than that of October. Rainfall is much the same from August to December (Table II). The only radical external change at the period of ovulation is the seasonal advent of insect food.

The young hatch late in November, December, or even in January. There is no second brood. Yet, unlike the testes of other single-brooded species (Rook, Hooded Crow,[181] Fulmar[166]) examined, those of the male Satin-bird do not immediately collapse and undergo post-nuptial tubule metamorphosis. It is instructive to compare the male and female gonads at this period (Pl. 8). Although the post-nuptial-metamorphosis of the seminiferous tubules has begun, and some necrotic sperm debris has appeared, the testis has remained large and functional long after the ovary has regressed and collapsed. It is suggested that the continuation of male display at the bower after the female has left to build her nest is associated with a prolonged hypophysial and gonadotrophic activity. We know that the injection of prolactin leads to sudden testis-collapse in some species.[142, 179, 202] As the male Satin-bird takes no part in incubation, it could be expected that no such third pituitary factor would operate while display continued. The male remains potentially polygamous.

The young leave the nest generally between the middle of December and early in January. Now a remarkable spectacle can be seen. The female, the young, and perhaps the young of previous years (or perhaps those of a polygamous union) join the blue male at his display-ground. Sometimes two, or perhaps more, blue males may appear there. This is probably the only period of the year when more than one blue male is seen at the principal bower at the same time. Display occurs. Then, within a week or so, painting and other display ceases. The bower is now apt to be wrecked, but whether this is done by the owner or, more probably, by now unopposed marauders, is not known. The bower is not permanently rebuilt. During this last phase of bower-display the male gonads are in full metamorphosis. In one such bird the larger testis measured $11 \cdot 6 \times 8$ mm. and weighed 295 mg. The second testis measured only 8×7 mm. and weighed $200 \cdot 5$ mg. Post-nuptial lipids had begun to appear in the tubules. Some of the tubules were reduced to a diameter of 140μ, and still contained large bunches of spermatozoa and many sperms free in the lumina along with accumulating debris. Other tubules were reduced to a diameter of about 100μ, and had their centres packed with a mass of metamorphosing material (Pl. 9 d). The interstitium had become invaded by numerous large blood-vessels, some measuring as much as 50μ in diameter. A few Leydig cells of the former generation remained among the new upsurge of juvenile cells that were everywhere apparent. A new testis tunic was in formation and measured 50μ in diameter.

The family group now joins others to form the big mobile and noisy flocks that begin to wander through the forest in search of fruit. Males taken from the flock late in February have completely metamorphosed (Pl. 9 a) and collapsed testes which are similar to those of other passerine birds during the period of interstitial and tubule exhaustion that prevails when the young first appear. Thus, in the Satin Bower-bird metamorphosis is merely delayed until the male ceases the display that he has carried on long past the period when the males of most other species have become sexually inactive and are helping with incubation and the care of the young.

Heredity and behaviour

It has been seen that very soon after the young leave the nest they become associated with the male parent, and possibly with other males, while the final displays of the season occur. Further, early in the following display season, the blue male sometimes allows young males at the bower-side (in captivity at least) to watch his gyrations. Therefore, it is certainly possible that all the manifold aspects of display could be learnt by the young males from the adults. This is the opinion held by naturalists who, however, have carried out no experiments to support their views.

No absolute proof is yet available, but the balance of evidence suggests that, contrary to the above view, display, bower-building, colour-collection, bower painting, and the capacity for vocal mimicry are expressions of innate behaviour patterns. These are automatically called into operation under the periodical influence of sex hormones. We have seen that gonadectomy inhibits normal display and that injections of testosterone re-establish it. At the same time it is probable that young birds improve their bower-building capacity with experience and practice. Certainly many bowers at which green birds communally display are very rudimentary compared with the bowers at which either solitary green or blue males display in front of a female.

It was unlucky that the young male bred by Hirst in 1937 was allowed to watch his sire at display before Hirst was told of the great experimental value of the bird. Thus, the young bird could have learnt its display by example. By a most remarkable piece of fortune, however, Hirst was comparatively unfamiliar with their habits in the bush. Therefore, although he went to considerable pains to provide his birds with growing vegetation and undisturbed breeding conditions, he did not give them coloured objects. So it was accidentally proved that, in the case of one pair of captive birds, reproduction took place after the construction of a bower and prolonged display, but without the employment of coloured display-things. The blue male displayed with merely a few odd twigs that he collected in the aviary. So it was that the green male offspring reached his fourth year, and had already acquired a mottling of blue feathers, without any personal experience of coloured display-things. Blue in particular he had never seen apart from his parents' eyes and the probable occasional glimpse of a passing blue frock or scarf. Thus the possibility that this young green male had learnt from its father how to discriminate colours and to assemble them at the bower was automatically and, up to the present time, uniquely eliminated. In his fourth year the young male was provided with the proportionally coloured cards used in the experiments mentioned previously. The results of six tests are so interesting that they are reported on in full:[115]

Card No. 1. ¾ Blue and ¼ Red. Selected on each occasion and carried to the bower, equalling a 100% result.

Card No. 2. ½ Blue and ½ Red. 3 times or 50%.

Card No. 3. ¾ Red and ¼ Blue. Once or 16·6% (approximately).
Card No. 4. ¾ Grey and ¼ Red. Nil.
Card No. 5. ½ Grey and ½ Red. Nil.
Card No. 6. ½ Green and ½ Red. Nil.
Card No. 7. ½ Yellow and ½ Red. Nil.
Card No. 8. ½ Blue and ½ Yellow. 4 times or 66·6% (approximately).
Card No. 9. ½ Blue and ½ Green. 5 times or 83·3% (approximately).

In the various tests the cards were quickly selected and carried together to the bower in the following colour combinations:

1st Test. Cards taken, Nos. 1, 2, 8, 9.
2nd Test. ,, Nos. 1, 9.
3rd Test. ,, Nos. 1, 2, 8.
4th Test. ,, Nos. 1, 2, 9.
5th Test. ,, Nos. 1, 3, 9, 8.
6th Test. ,, Nos. 1, 8, 9.

The placement of the cards in the different tests was varied as follows:

1st Test. Scattered.
2nd Test. In row not in numerical order.
3rd Test. In row in numerical order.
4th Test. Reversed in numerical order.
5th Test. In form of square.
6th Test. In form of square but with colour arrangement varied.

From the above three fundamental facts emerged:
1. Almost certainly, the urge of the young male Satin-bird to decorate its display-ground with coloured objects is innate and not learnt by watching similar activity by older males.
2. The young male Satin-bird, as was indicated by previous experiments and observations on old males, is much more conservative in its choice of greenish-yellow and grey than of blue. The greenish-yellow and grey shades on the cards closely approximated those of the flower *Billardiera scandens* and puff-ball fungus respectively, but the bird rejected them when they were not associated with blue.
3. The Satin-bird has no aversion to bright red provided it is associated with blue, a colour that it normally collects with avidity.

Some additional evidence that display is innate will be provided in a later chapter on the Spotted Bower-bird.

At the same time it must be remembered that the young of some passerine species appear to have their characteristic song imprinted[108, 153] after hearing the song during the first season in which gonad development takes place.[222] We have at present no knowledge of the internal state of recently hatched bower-birds, but there is evidence that very young individuals of other species undergo activities of a seemingly sexual nature. Even if the young Satin-birds are too immature to learn during the brief post-nuptial phase of bower-display, they undoubtedly see the adults displaying in flock a little later, and at the communal sub-bowers and platforms during subsequent seasons. At Bart's we are now trying to breed Satin-birds so that the young can be isolated early from their sires and given the opportunity to display without any possibility of parental example. It is unfortunate that bower-birds are among the most difficult of all species to breed in captivity.

Discussion

We have seen that the gonad cycle of the Satin Bower-bird follows a very similar course to those of many less remarkable species, the only notable difference being the prolongation of the period of spermatogenesis, including its final extension while the male continues his display during the period when the female alone builds her nest and tends the eggs and young. It is worth while, even at the risk of a little repetition, briefly to recapitulate the behaviour sequence.

Pair-bonds seem to be formed in the noisy autumn flocks after the post-nuptial regeneration of the Leydig cells of the interstitium at a period of more or less continuous arboreal, and spasmodic terrestrial, bower-display. At the period when the Leydig cells exhibit a seasonal rearrangement and the seminiferous tubules become clear of lipoidal debris, the male permanently leaves the flock, builds a bower, collects display-things, and brings to his territory a female. The primary sex organs of both birds develop immediately, and those of the male very considerably. Without doubt sex hormones are liberated and these seemingly act also in the central nervous system. Little is known about the localization of innate mechanisms in birds, although it is generally considered that the remarkably developed corpus striatum is the seat of many of them.[239] Sex centres occur within the tuber cinereum.[139] It is still true to say, however, that the question of the neural control of avian behaviour patterns is an almost totally uninvestigated field. All we really know is that sex hormones are carried to the brain and there act upon specialized centres and call into play hereditary behaviour patterns located therein (Fig. 3). The urge to build a bower and to display appears to be in its way as automatic as hunger. During the display season blue birds will accept blue objects in preference to food if they are not hungry, and in squalid private aviaries I have seen lone males persist in building and rebuilding a meagre bower with any odd twigs that they can find even when the structure is repeatedly pulled down by parrots or other birds. So intense is the feeling for blue that one captive blue male is known to have killed blue Cuban Finches and placed them dead on his display-ground. The same bird killed a blue-headed adult male African Nonpareil Finch a little later. The owner of the aviary at first imagined that the finches died from natural causes. He replaced the Nonpareil Finch with another. The second finch, and later a third, were found dead on the display-ground. 'The last time I released the bird, the "Satin" came like a lightning flash and smashed him in front of me', wrote the astonished owner.[200]

The blue male postures energetically in front of the female during the weeks following her arrival at the bower. This activity helps cement the pair-bond. Male and female have a strong mutual attraction for each other at this period, and there seems no doubt that the presence of the other has its effect on the internal physiology of both. The male displays violently with a choice from the large variety of coloured objects that he has collected. His display has within it the elements of a displaced combat drive. He frequently assumes what appears to be an angry attitude, but this is directed towards a blue feather or brown snail-shell or other object held in his beak. One of the male's postures suggests coition. He expends tremendous physical energy in display, frequently presenting his flashing surface to the mate, who watches intently. From favourite high perches above his bower-territory he mimics other birds, calls loudly with a long ringing note and makes flights from tree to tree, and so proclaims his domination over the immediate area. However, if he ventures from his territory in search of food his bower may be wrecked, and his display-ground is often robbed of its decorations by a neighbouring rival who comes skulking through the undergrowth, but who flees immediately when the owner comes swooping through the timber. The intruder never stops to fight. The flashing dark plumage of the adult male acts as both a threat and a warning to enemies and competitors: it is aposematic and episematic in the classification of Huxley.[121]

The respective behaviour of marauder and defender is thus broadly similar to the characteristic pattern exhibited by more commonplace species when their territory is invaded by rivals. The swift reaction of the bower-owner depends on at least a three-fold complex of primary factors. He is (1) faced with an environmental situation to which (2) an innate releasing mechanism[153, 274] is attuned to respond if (3) certain seasonally recurring internal factors are operating at the moment. If the blue bird meets a rival away from his bower-territory he will not seek combat. It is only the sight of a rival near the blue male's own display-ground that releases the latent aggression that is waiting to be unleashed in such a specific situation. If a stuffed blue male is placed on the display-ground he strikes it violently, sending it sprawling into the undergrowth. However, internal stimuli are equally important. If the gonads are not enlarged and highly active, the gestalt does not evoke the violent reaction. The male will tolerate others of his kind and colour at the bower in January and February at the time of testis metamorphosis immediately after the breeding season. It would seem, then, that an environmental situation, an innate behaviour pattern, and an appropriate hormone threshold are all necessary for the simple attacking reaction. The brain, of course, is a highly vascular organ, bathed in fluid, and is as much affected by hormones as are the accessory sexual organs which are so demonstrably altered at appropriate times of the year. It is almost certain too that the sympatho-adrenal system is involved.

The marauding bird is faced with an entirely different environmental pattern. An unguarded display-ground, the focal point of his rival's sexual activity, releases in him a drive to wreck and to rob. It is interesting that the urge to carry away the display-things (coloured somewhat like the female) seems to be secondary. He appears to take the display-things only after wrecking is completed, or hastily snatches them up as their owner appears. The interloper is unsure of himself on the other's display-ground. While in alien territory he works quietly and with speed. Immediately the dark bulk of the owner hurtles into sight the raider snatches a decoration and retreats. Sex hormones, are, of course, present in his own blood-stream; in fact, they probably brought him to the rival's bower in the first place. But, interrupted on another's territory, his whole emphasis is on rapid flight. The sympatho-adrenal system is again no doubt involved but, in the interloper, to a diametrically opposite purpose. Thus we have in the Satin-bird a remarkably neat example of how different environmental situations can call into play different behaviour patterns in two individuals which, it is fair to assume, are virtually identical in their seasonal physiology.

The brilliant blue eyes of the male probably serve a short-range threat function. When the bird is excited the eyes take on a bulging, glaring appearance which probably has an alarming effect on rivals. These brilliant characters are reinforced by the aggressive temperament, manifesting itself in harsh 'snarling' utterances, when any enemy, or intruder of other large species, ventures near its bower. Some of the bower activities recall Hingston's[111] controversial display-threat hypothesis which holds that all significant animal colour subserves two main functions, concealment or threat. Broadly speaking Hingston disallows the epigamic or sexual display function of colour and, in the cases in which he admits that it may operate sexually, declares that its primary function is one of threat towards the opposite sex. 'The gesture made before the female is exactly the same gesture as that made before the rival in battle', he says, and 'courtship behaviour is hostile behaviour.' Huxley[121] has satisfactorily disposed of this argument as applied indiscriminately to bird courtship and display. At the same time, the 'fierce' intensity of the male blue bird at the bower, and the 'threatening' attitudes that he also adopts there, make it appear that in this species display is basically closely allied to the general threat complex which the bird has developed as part of the protection of its bower, display-things, and its territory. The watching female, of course, does not exhibit fear. The 'threat' gestures are directed at the display-things, not at her. What appear to be basically similar gestures, however, exercise a deterring effect upon rival males. It would seem that gestures originally evolved as part of a mechanism of aggression have become part of a display mechanism which is released during emotional excitement of a somewhat different kind.

Although the female has developing oocytes, and is usually present at the bower, the physical energy of the male is for many weeks directed only toward the display-things even though he has achieved spermatogenesis and is, physically speaking, ready for reproduction. The display-things which he flings about so violently match fairly closely her colours. It is impossible to ignore the possibility that, during the long period in which the male is physically capable of reproduction, the display-objects, coloured in the female's image so to speak, serve to satisfy his physical appetite until the female is ready for, and probably invites, copulation. If this conclusion is accepted, a superficial anthropomorphic view of the phenomenon could suggest that the male's preoccupation with display-things is a form of sexual fetishism. This, however, would be incorrect. Fetishism is, of course, a human perversion whereby sexual excitement is unnaturally aroused by the use of inanimate objects of an astonishing variety chosen according to the abnormal and conditioned desires of the individual. Although one pair of captive Satin-birds reproduced several times in the absence of these coloured display-things, there is equally good evidence that the desire to gather and to display with such objects is perfectly normal, innate, and common to all males of the species, whatever their age and situation. The older male's appetite for blue is so great that it will continue to collect blue articles even after its display-ground has been experimentally smothered with them.

One could suggest an alternative hypothesis that the male chooses display-things in the image of rival *males* and that the violent ritual with, and at, them is partly a reflection of his own fierce antagonism towards competitors that try to invade his territory. Certainly the chosen colours match equally well those of young green males. In particular, the peculiar favoured shade of greenish-yellow matches the beak-tips of blue adult males more than it does any prominent female character. Yet apart from odd yellow-vaned, black, moulted feathers, dark and reflective objects resembling the adult plumage are never used.*

* See Addenda, note 'B', p. 190.

In either case the collection heightens reproductive efficiency. It assists the male in display to the female during the months in which he must retain her interest if she is to be with him when ready for insemination.

The bizarre habit of bower-painting is probably an extension of, and possibly a substitution for, the courtship-feeding that is practised by so many birds during the sexual season. It is legitimate tentatively to suppose that the plastering of the bower walls with masticated fruit-pulp is a substitution activity in that it enables the bird to satisfy its innate drives towards courtship-feeding without actual contact with the watching female at the bower.

It will be recalled that almost all of the fruits so far observed to be used in bower-painting are bluish in colour. It is possible, of course, that this is pure coincidence. A remarkable elaboration of the painting habit, that of mixing charcoal and saliva and, with the help of a manufactured wad of fibre, the achievement of an embellishing black band inside the bower, may be an aesthetic extension of a basic drive whose evolution has been permitted by the prolonged period of sex hormone liberation while the male awaits the female response.

It has been suggested to me in conversation in Europe that the charcoal-painting elaboration is of necessity as recent as the introduction of fire by man. This is almost certainly not a fact. Lightning not infrequently starts bush fires in dry countries. A tree struck by lightning can lead to the devastation of hundreds of square miles of grassy *Eucalyptus* forest and the charred logs which result survive for a great number of years. It is unlikely that Satin-birds inhabiting the savannah fringes of the south-eastern Australian rain-forests have ever been without an adequate supply of charcoal.*

We have seen that while the gonads are maturing, and testosterone is being liberated, the male's physical attention is canalized into violent, and sometimes almost frenzied, activity directed at the display-things. Here we seem to have a dynamic metamorphosis of the sexual drive. If we were not careful to distinguish a normal, innate and cyclical avian behaviourism from a socially conditioned, psychical human aberration, we might be tempted to identify the Satin-bird's behaviour with sexual sublimation or conversion in the Freudian sense. The excitable bower activity is essentially sexual activity which may be the result of a temporary sublimation of a highly specialized, but perfectly normal sort.[173] The prolonged period of sex-hormone liberation and the aloof presence of the watching female has probably led to the development of the highly refined degree of ritualization, bower-painting, and so on. These activities have flowed into the vacuum, so to speak, created by the

* See Addenda, note 'C', p. 192.

female's unwillingness (because of the, as yet, unsatisfactory state of the environment for breeding) to exhibit the sign stimulus to which an innate releasing mechanism in the male is probably hereditarily attuned to respond. It is probable that the exciting activities at the bower have become highly enjoyable in themselves and so, after the post-nuptial gonad rehabilitation has proceeded to a certain degree, the males spasmodically return to their territories to display and, as the external conditions become more favourable, perhaps annex females and territories earlier in the season than is dictated by strict reproductive utility. Thus, some individuals occasionally bring the female to their territories about 4 months before the fertilization date and 2 or 3 months before the majority do so.

If we agree that the remarkable elaboration of the display-pattern has been made possible by a temporary sublimation, during prolonged androgen liberation, of the primary aim of the male sexual impulse, viz. the insemination of the female, it still must not be confused with sexual sublimation such as occurs in man. There is a vast difference between the collection of cats or postage stamps, or the over-emphasized, sometimes savage, religiosity of the mildly or sometimes highly psychotic unfulfilled solitary human and the regular seasonal development in every member of an avian species of an elaborate behaviour form which, practised in the mate's presence, demonstrably assists him to retain her interest and helps ensure that fertilization will occur at the most appropriate season of the year. The avian activity is a means towards, not a substitute for, successful reproduction. It is perhaps scarcely necessary to stress this difference, but as new discoveries relating to bower-birds have almost always provoked a flood of anthropomorphic generalization, it seems not unwise to do so. There has, for example, been a recent reference in a medical journal[4] to 'sexual fetishism' in bower-birds in connexion with their predilection for inanimate coloured objects during display.

Along with his flashing eyes and plumage (Pl. 1), the male presents to the watching female a vocal accompaniment of remarkable tone and rhythm. One writer has described this succession of notes as resembling 'the noise made by small rapidly running cog-wheels, accompanied by a deep hissing sound'. And again, as 'peculiar machinery-working-noises'.[205] Another[42] has written of these notes as sounding 'as if a series of gears were running inside the bird'.

Display activity is accompanied by the slow maturation of the female gonad. Thus, although it no doubt helps synchronize the reproductive processes of the pair, it apparently has no great accelerating effect on the female sexual cycle. The bower is so exclusively the possession of the male that once a female becomes attached to it she will in all probability be ultimately inseminated by its domineering owner. Early in the season, before the pair-bond is fully established, the male will not display fully unless the female is present. Later, when the partnership is adequately formed and the female oocytes have enlarged, the male permits her prolonged absences and carries on unconcernedly with his bower-renovation, colour-collecting, and even display. Some naturalists who have been most aware of this phase of the reproductive cycle have declared that display and bower-building are essentially relaxative 'hobbies', and, as such, now only vaguely connected with the primeval impulse of sex.

We do not yet know what precise combination of environmental stimuli causes the female innate releasing mechanism to dictate a special posture or other activity that will invite the transference of the male's physical attention from his coloured display-objects to the similarly coloured female herself. At present we know only that fertilization is delayed for months until a time when the forest becomes full of the flying insects on which the young are fed. We know, too, that barred from protein food by captivity, the female is unable to sustain her chicks and that they die of malnutrition during the first fortnight of life.

Much has been written of the alleged high degree of intelligence of bower-birds. Some naturalists have, somewhat lyrically, credited them with conscious aestheticism and with the power, even, of deductive reasoning. There is no evidence whatever that they are more intelligent than any other crow-like bird. I have observed that the Satin-bird is neither sufficiently intelligent to scratch aside leaves under which a potential prey has run, nor to attempt to reach an animal that is clearly visible in a cocoon that was formed naturally in its aviary. The Stage-maker or Tooth-billed 'Bower-bird', as will be shown in Chapter 21, has not learned that certain display-leaves, which it spends minutes detaching with its extraordinary 'toothed' bill, could be almost instantly freed with a single downward thrust of the beak.

The Satin-bird can undoubtedly sometimes learn quickly by experience, but so, too, can other birds. While engaged in cinematographic photography of a blue bird at its bower, Chaffer and I first briefly accustomed the animal to a mechanical noise by the operation of an egg-whisk. This avoided the waste of many feet of film. One year later, at a sub-bower about a hundred yards away, further cine-photography was attempted by Ramsay and me. The green birds fled immediately. The blue bird started sharply and then carried on, unafraid, with his activity. It seems legitimate to conclude that he remembered the mechanical sound of one year previously, recognized it as harmless, and acted in the light of previous experience.

It is necessary to consider the striking dimorphism of the Satin-bird and the possibility of sexual selection and polygamy. Females and young males are greenish. The males first become slightly mottled with individual dark feathers. It is not until the following moult that they become wholly blue in coloration. But before they have grown a single blue feather, young males build full-scale bowers, annex females, achieve spermatogeneses, and sometimes breed. The castration of blue birds does not cause a return of the former green plumage. Nor does it halt colour-change in mottled birds. It is obvious, therefore, that such change is not under the control of sex hormones. There is, incidentally, a record[47] of the dissection of a blue bird which contained oocytes. Birds not infrequently alter sex owing to various pathological changes involving the sex glands. The most likely explanation of the presence of an ovary in the blue bird mentioned above is that it underwent sex reversal without a corresponding change of plumage colour.

Green males compete with each other in their endeavours to acquire and to retain coloured objects, and observation of occasional combats in the aviaries suggested that although their plumage pattern is essentially cryptic or concealing, the prominent blue eyes and bright yellow underwing feathers (with contrasting grey terminations) serve, among themselves, an aposematic or threat function. When green Satin-birds fight, they do not stand up to each other like, for example, a pair of domestic cocks. They tumble over, and most of the combat takes place with the birds on their backs or sides, pecking at the other's eyes or striking out ineffectually with the feet. In these combats the bright yellow underwing pattern is conspicuous, and no doubt, along with the eyes and other characters, exerts a deterring effect on the antagonist. In most species, bird warfare is psychological warfare. Combat is generally a comparatively harmless affair and usually amounts to little more than a token display from which the less aggressive individual retreats. Among green bower-birds (in captivity) a curious submission reflex appears to operate as soon as one bird gets the better of another. The vanquished bird rolls over in an attitude of surrender and the victor, after a slight pause, hops away. Huxley suggests that this mode of behaviour is common in many gregarious species, including mammals (e.g. dog). It is the converse of dominance threat. 'Just as it is to the advantage of the more powerful individual to establish his (or her) dominance without having actually to fight for it, so it is to the advantage of the more helpless individual to establish his peaceful intentions without having to suffer actual aggression.'[121]

Blue birds, on the other hand, have been known fatally to injure green birds which could not escape their sustained savagery during the period of androgen liberation and sexual aggression. Hirst's[115] blue male adopted an attitude of 'tolerance without affection' towards its male offspring until the second breeding season after hatching. Then it viciously attacked the young bird. The attacks waned when the female left the bower to nest. The young bird was left in peace until the following season when the attacks were renewed and the young bird had to be removed in order to save its life. It was apparently quite incapable of defending itself and recovered health only after several days. Then it built a full-sized bower of its own and sparred with its sire through the protective wire.

Although each blue male has only one female in attendance at the principal bower once the sexual season is well under way, it is possible that his frequent visits to the sub-bowers and run-ways result, by virtue of his plumage and domineering temperament, in the fertilization of a second or even more females there. It will be recalled that the testes of the blue Satin-bird do not immediately metamorphose after the female has been inseminated but, unlike those of many other species, remain in a productive condition. On the other hand, it must be remembered that on each occasion that a blue male was removed from his bower, it and the female were taken over not by a neighbouring bower-building blue bird, but by a male green bird which appeared to be unmated, territorially unattached and therefore free to do so.

Whether the species is polygamous or not, the sub-bowers and run-ways still have the useful function of aggregating unmated young green birds of both sexes at a central place, where the males exercise their innate capacity for building, colour selection, and display. It is probable that there is a tendency for the previous offspring to display at a sub-bower near the principal bower of their sire. There is, however, a considerable migration between the sub-bowers, for near one often as many as seventeen green birds have been observed together. Should a bower-building blue, or a senior green, male lose his mate it would seem that a second young female is always available at a sub-bower or arena nearby.

Thus, a form of sexual selection probably occurs. Although the pair-bond has been at least partially formed before the male brings the female to the bower, it seems probable that a female stays with a male only if he and his superior display continue to stimulate her to do so. It will be remembered that when a female is removed experimentally from a bower the male brings another female to it within a few days.

Today we know that sexual selection, as propounded by Darwin,[64] is a rare phenomenon. However, in some gregarious, polygamous species (e.g. European Ruff, Blackcock) it undeniably exists. Darwin hypothesized a struggle between males for mates and the probability of female choice between rivals which, by virtue of their colours, plumes, ruffs, crests, or other special structures of display and/or combat, would succeed in joining their gametes with those of the discriminating females. This indeed may occur in the Satin Bower-bird. It is difficult to believe that the astonishing display elaborations of the male Satin-bird give him no greater number of mates, and probably a less number of offspring (one or two, rarely three) than are achieved by the most drab fly-catcher or scrub-wren sharing his territory: or, even more appositely, no greater number of mates or offspring than are got by the green males which are dominated by the blue birds, yet which sometimes breed. The territorial domination, the bower-building and precise choice of coloured display-things, the elaborate ritual, threat, epigamic coloration, the noise —these would all suggest a capacity to gain a plurality of females, if so desired, against younger green birds, or ageing blue birds, or those with the above characters less efficiently displayed. (See Addenda, note 'D', p. 192.)

The display of the Satin Bower-bird may be perhaps compared with the gigantic size and weapon development of certain Mesozoic reptiles in which it has been suggested that evolutionary process may have 'acquired a momentum which took it past the point at which it would have ceased on a basis of utility'. Haldane[107] has further suggested that such specializations (often in the case of the male only) have sometimes been a prelude to extinction, and that creatures have 'literally sunk under the weight of their own armaments'. During the period when his testes are at their maximum seasonal development, the male Satin-bird, even though unaccustomed to human observation, will perform fearlessly when he knows that the partially concealed human is present. It immediately occurs to mind that the male's intense preoccupation with the bower should place him at a disadvantage with marsupial 'cats' (*Dasyurus*), dingos, imported foxes, and domestic cats that have gone wild. However, the Satin-bird remains fairly plentiful, despite the male's noisy display and boldness at the conspicuous display-ground. It shows no sign of disappearing under the weight of its social armaments.

Again, it might be suggested that if the environment changed in some way that prevented the species from carrying out the whole complex sequence of its display, it might disappear before it could adapt itself to a less elaborate technique of reproduction. Apparently such is not the case. Satin-birds which do not paint their bowers reproduce efficiently. Further, deprived of coloured display-things in captivity a pair bred several years in succession. We cannot say whether a pair could reproduce if prevented from building a bower. In nature, of course, such a contingency—or the lack of coloured display-objects for that matter—could not arise.

It has been claimed that Satin-birds 'die off'[144] shortly after they assume the complete blue plumage. There is no evidence that this is true. One blue captive control bird lived for a decade, making a probable total age of at least 14 years. No data concerning the longevity of bush birds are available. But in view of the delayed assumption of the male adult plumage, the tremendous breeding advantage that it confers, and the small clutch size (one, usually two, and rarely three eggs), it would be surprising if Satin-birds died soon after they became blue. If predator or other pressure made for short life we should expect an adaptation towards a greater seasonal egg-production. It could be argued that day-length at the time of hatching might be too short to enable the female to gather enough food to sustain a greater number of chicks, or that the external temperature might be too low to allow the young to survive her prolonged absences from the nest. In view of the unvaryingly high temperatures (63–67° F.) of November and December and the seasonal plenitude of insect food it would be hard to defend such a thesis.

The habits of the Satin-bird make it difficult, or perhaps impossible, to attack the matter experimentally by adding to a nest young taken from a second one and so to observe whether the female could successfully rear four or five chicks single-handed. The impression is gained that there exists in this, and other bower-birds, some clutch-limiting mechanism about whose mode of operation we have at present no knowledge whatever.

No precise data are available, but it would seem that the environment, with its great early fruit-harvest and later enormous insect productivity, at all times contains far greater food supplies than are needed by the Satin-bird and its economic competitors. Predator-pressure, as a matter of fact, appears to be extremely low once the Satin-bird leaves the nest, although eggs and young are not infrequently taken by goannas (*Varanus* spp.). The pioneer collector Jackson[131] testified to this in sober Edwardian prose. Jackson climbed to a Satin-bird's nest in order to steal its eggs and found that a 4-foot-long lizard, with the same intention, had arrived there first. 'The ugly creature,' he said, 'in its sudden amazement, jumped on my head, and then descended to the ground. The feeling to me was very unpleasant, and of rather a rare nature.'

In the Satin-bird we have the impression that strenuous intra-sexual competition, coupled possibly with polygamy, has resulted in the evolution, by means of natural selection, of the remarkable eyes, plumage, aggressive temperament, and the other elaborate display specializations that we see today. It is a remarkable fact that the evolution of this astonishingly complex and, to some degree, aesthetic reproductive mechanism has apparently rendered the species neither more nor less numerically successful than many other quite undistinguished birds of its immediate environment.

5

SPOTTED BOWER-BIRD
Chlamydera maculata (Gould)

THE strange habits of the Spotted Bower-bird (Pl. 10), rather than the generally cryptic plumage of the bird itself, compelled to it the attention of colonists soon after the crossing of the Blue Mountains and the settlement of the Western Plains of New South Wales. Before Gould came to Australia he found specimens of unknown origin from 'New Holland' in the collection of the Zoological Society of London. He described[100] the species. On arrival in Sydney he found more skins and a bower preserved in the Australian Museum.

The Spotted Bower-bird is a generally brownish bird, mottled with rufous and golden buff. This cryptic and harmonious colour pattern is enlivened by small erectile mantle or neck-frill (Pl. 11 a) which is generally described as pink or rose-lilac. This crest is conspicuous only when it is raised. It is an epigamic feature that takes on a silvery iridescence when erected and displayed in sunshine. Comparison of the specialized nuchal feathers with Ridgway's[231] colour standards suggests that the colour alternates between something approaching 'rose-purple' and 'thurlite pink', whilst from other aspects the mantle feathers are not unlike 'bittersweet orange' in coloration. Microscopic examination reveals that the rose-lilac colour is a physical phenomenon caused by the modification of white light by the specialized cellular structure of the greatly elongated distal barbs of the erectile feathers which appear to be yellow or orange in coloration. Pl. 11 b shows the differential structure of the coloured erectile feathers and the drab ordinary nape feathers that surround them. On the distal barbs of the erectile feathers no barbules occur. Pl. 11 c illustrates a distal part of an individual barb and the cellular arrangement that is responsible for the alteration of light. Pl. 11 d reveals this pattern under high power showing the arrangement of individual cells and their nuclei. Pl. 11 e shows a barb closer to its base. Here only odd cells of the specialized kind occur, and likewise only odd points of altered light appear under the microscope when the reflector is moved and the light redirected about the barb. Manipulation of the reflector, and the use of ordinary white light, will cause a change of colour from yellow (centre of barb) to a most beautiful lilac (at margin), both colours occurring at different parts of the barb at once. Yellow and lilac, of course, are complementary colours. It is clear, then, that the play of colour of the erectile nuchal mantle of the Spotted Bower-bird is governed by the microscopic cellular arrangement of the distal portion of the barbs of its component feathers. Further, when the bird moves in the sunshine the lilac crest takes on a silver iridescence which is, in a limited way, comparable to the glistening plumage of the male Satin-bird.

The Spotted Bower-bird is about 12 inches long. The sexes, when adult, appear to be almost identical externally, though there have been suggestions that the neck-frill is better developed in the male. Young birds, besides having a more heavily marked throat and chest, lack the crest altogether. It is not yet known at what age it appears. The species feeds on fruit, including the berries of the white cedar (*Melia azedarach*) which in northern New South Wales are believed to be poisonous to stock and are said to have killed children.*[270] Insects are also eaten.[143] The Spotted Bower-bird is nowhere very common but it ranges widely through the scrubby, comparatively low rainfall areas† of the Australian mainland where it is not inhibited by deserts or replaced in the north by its slightly larger congener, the Great Grey Bower-bird (*C. nuchalis*) (Fig. 7).

Chlamydera m. guttata replaces *C. m. maculata* in the more central and western parts of the continent. The western Spotted Bower-bird is darker and more conspicuously spotted (especially in the collar region) and is consequently the handsomer bird. The two forms are separated by a considerable gap (Fig. 7) and may already constitute different species. No general agreement regarding this has been reached and so both forms are treated together in the present account. The precise range of the species is still unknown. Serventy and Whittell[254] say that in Western Australia the distribution of *C. m. guttata* is determined largely by that of a native fig (*Ficus platypoda*) and this may be true of the species in Central Australia as well. This remarkable fig manages to live in barren and periodically waterless areas where its roots penetrate small fissures in granite and other outcrops and somehow find sustenance. It provides fruit for the Spotted Bower-bird, and sometimes shelter for its bower. Bowers are never found far from water. In areas where natural springs or other more or less permanent water are absent a bower may be built near

---

* Several Australian birds eat with impunity fruits that kill mammals. It might be expected that they somehow detoxicate the poisonous principle but apparently this is not so. If dogs or cats eat the *bones* of Bronze-wing Pigeons which have eaten the seeds of the heart-leaved poison (*Gastrolobium bilobum*), one of the West Australian Leguminosae, they die after convulsions and paralysis. The flesh of the pigeons is harmless.[285] Here we seem to have a remarkable aspect of drought adaptation that enables birds to take advantage of all possible food in relatively unsustenable areas.

† During the severe inland drought of 1902 a stray Spotted Bower-bird was collected near Sydney within 10 miles of the coast. The skin is in the Australian Museum, Sydney.

74 SPOTTED BOWER-BIRD

a windmill. Sometimes one is built within full sight of a homestead veranda.

The display-ground, the bower and its painting

The bower is a simple double-walled structure (Pl. 10) roughly similar in architecture to that of the Satin-bird and all other Australian

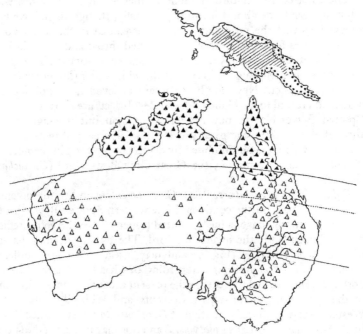

FIG. 7. Approximate distribution of Genus *Chlamydera*.
White triangles: Spotted Bower-bird (*C. maculata*).
Black triangles: Great Grey Bower-bird (*C. nuchalis*).
Spots: Fawn-breasted Bower-bird (*C. cerviniventris*).
Stripes: Possible distribution of Yellow-breasted Bower-bird (*C. lauterbachi*). (Both New Guinea ranges are purely provisional.)

avenue-builders (Fig. 1). The outer part of each wall is generally composed of thin twigs and the inner part of grass-stems. The walls are 5 to 9 inches thick. They are 6 to 9 inches apart, stand between 10 and 20 inches high and are 15 to 30 inches long. North,[205] however, describes deviations from the common type and has a report that one Central Australian bower (*C. m. guttata*) measured about 7 feet long. McGilp[192] too has described a desert bower (of *C. m. guttata*) which was 3 feet 5 inches long and was constructed entirely of the wire-like stems of grass.

There is a reliable description of a West Australian display-ground which lacked the usual surmounting bower. Whitlock[281] described 'a large number of small sticks which had been carried to a clear space, with a feather or two and a few sandal-wood nuts'. Whitlock's own persuasive calls brought seven birds to the vicinity. One of these indulged for about an hour in what seems to be the normal nuptial display to a female. The remaining birds sat around watching the human intruder. The period was September. This suggests that the Spotted Bower-bird sometimes builds a rudimentary display-ground similar to those that are the property of young male Satin-birds. Another writer[80] describes a platform or runway measuring 20 by 16 inches built on a horizontal branch of a gum-tree high above the ground. Although sticks were added from time to time, no display was seen there. Among normal bowers a considerable degree of variation is manifest. Even at the same time of the season some are obviously better built and cared for than others.

The evidence concerning bower-orientation is conflicting. It will be recalled that almost every mature Satin Bower-bird builds its bower across the path of the rising sun and the same habit will be described in the Great Grey Bower-bird. The present species, *C. maculata*, shows no such constancy. Robinson[238] says that West Australian bowers may face any direction. Gaukrodger[88] says that 80 per cent. of central Queensland bowers were built *along* the sun's path, within a few points of due east and west. Jackson[129] took bearings on six northern New South Wales bowers and found that half pointed east and west and the remainder north and south. A resident[87] of St. George, southern Queensland, was kind enough especially to take bearings on the five bowers near his home. He reported that all except one pointed within 20° of north and south. The exceptional bower was orientated at 327° (33° west of north). Of two other southern Queensland bowers one was orientated at almost north and south, and the second absolutely north-west and south-east.[74] At Quantambone another bower was examined on a bearing of 8°. Sullivan[270] has described a double bower in which the two structures were almost touching and placed at an angle of about 70° from each other.

The bower is generally built beneath trees whose branches are sufficiently low (often touching the ground or rocks) to provide concealment. Neither bird nor bower is conspicuous at a very short distance. It is generally the noise made by the birds themselves that leads to the discovery of their display-ground. Bower and display-ground usually cover an irregular area of about 6 feet long. The bower is built approximately in the centre of this area and at each entrance, or a little to one side, is a large collection of display-things. Gould[102] estimated that about half a bushel of bones and bivalve shells decorated each bower entrance and noted that the 'crania of small mammalia' and other bones were always those that were 'bleached by exposure to the rays of the sun or the campfires of the natives'. He noted that smooth water-worn stones were also used, and observed that as the bowers were often built at considerable distances from the rivers, their transport to the display-ground must be a task of great labour for the birds. Nowadays, since the introduction of sheep and rabbits, the chief decorations are great quantities of the whitened bones of these animals, particularly the vertebrae of the former. McLennan[193] counted 1,320 bones on one Victorian bower. Along with the bones are found bleached eggshells and crab-carapaces, but green seed-pods, pine-cones, berries, and seeds are found as well. McGilp estimated that there were between 200 and 300 land-shells at one end of the desert bower mentioned previously. At the other entrance the bird had constructed a platform of 'rather large sticks' which was about 6 feet in diameter and from 5 to 11 inches deep. The bird had evidently built up this stage until it was on the same level as the bower and the 'shell-depot', for the ground dipped suddenly away at one side of the bower. The whole edifice—bower and display-ground—covered an area 13 feet 8 inches long. Upon the raised platform were some black stones (burnt by fire) and bleached bones. No shells were exhibited at this end of the bower, but sunk in the twigs of the platform were many shells and bones that were no longer used. Outside the display-ground were a few green berries and figs. The whole great structure was completely hidden by the drooping branches of a silver wattle. Whitlock[282] has given a description of decorations of a bower that he found far from white habitation at an altitude of 2,000 feet in the Hamersley Ranges, in October. The principal display-things, accumulated at the northern end of the bower, were small flat stones, up to about an inch in length, either white or slate-grey in colour. Other articles were the white buds of the moon-flower (*Capparis*). This flower, incidentally, opens during the night, giving off a sweet perfume, but is unable to withstand the heat of the day. 'The playground also contained green pods, almost like that of the garden pea. . . .' A few *Acacia* seeds in the avenue completed the decorations.

With the advent of European settlement the Spotted Bower-bird has come to add an astonishing assortment of materials. Bones still predominate but to these are now added broken glass, brass cartridge cases, pieces of tin, lead, zinc and nails, screws, bolts, and wire. North says that 'scissors, knives, spoons, forks, thimbles, coins' have been pilfered from houses and tents. Jewellery is occasionally found on bowers.[36, 47, 205] On one collecting trip North found the birds difficult to approach, but when he remained quiet they came to 'carefully examine the buttons on one's clothes'. Jackson reported that in his camps 'anything bright' had to be kept in a box; he was compelled to retrieve stolen table-ware and a pair of entomological forceps from a bower near his camp. Barratt[19] reported the removal by Spotted Bower-birds of a number of metal staples left along a line of post-holes that had been newly sunk. Another writer[9] records the disappearance of the ignition keys from a motor-car. Familiar with the habits of the Spotted Bower-bird the owner walked half a mile to the nearest display-ground and recovered the missing keys. Chisholm recounts a Spotted Bower-bird story involving the theft of a bushman's glass eye. Such an account would seem wildly improbable if we did not know something of the scavenging tendencies of *Chlamydera maculata*. The gentleman who owned the eye used to keep it in a cup of water overnight and he took a low view of its disappearance. The camp dog was suspected. 'The "bereaved" bushman waxed irate. . . . He cursed that dog; he kicked it briskly; he even proposed conducting a *post-mortem* examination upon it.' A fortnight later one of the men wanted to shoot a brumby stallion (wild horse) and having insufficient lead, visited a nearby bower in the expectation of finding some. He found not only the desired lead but his mate's eye as well, staring vacantly up from among the bones and shells on the display-ground![46] Highly reflective broken glass of pale green, white, amber, and mauve coloration is often brought to the bower, but clear greens, reds, yellows, or blues are rarely or never used under normal conditions.

In its choice of coloured, as distinct from reflecting, objects, the Spotted Bower-bird is also extremely conservative. We have seen that the Satin-bird will ignore a piece of onion until it decomposes to a satisfactory yellowish hue and then it will take it to the display-ground. The Spotted Bower-bird will take to its display-ground green chillies, but will discard them as soon as they begin to ripen to red.[88] Jarman[133a] reports that when eighteen yellow flowers were placed inside the bower of the western sub-species the owner removed them, one by one, 'with every indication of distaste'. Two stones, coloured with washing blue, and a piece of silver paper, were also carried right away from the structure. However, a shilling piece, placed at the entrance, was picked up and added to the collection within the bower.

On the other hand, one naturalist[80] asserts that Victorian birds brought proffered blue and red cartridge cases to a display-ground in that order. Yellow cases were ignored, and were removed by the birds when placed on the display-ground. Another[73] has claimed that the bird chooses amethyst-coloured glass fragments, 'in harmony with the bird's colour patch', but this has not been confirmed, and Jackson specifically states that there is a lilac 'everlasting-flower' (*Ptilotus exaltatus*), which blooms prolifically during the display season, yet the birds did not bring it to their bowers. Most authors have reported that the bird carelessly strews the display-objects at each entrance of the bower, but a group arrangement has been described in both sub-species[42, 192, 283] whereby bones were mostly placed separately from the other principal decorations. The western Spotted Bower-bird appears often to employ a greater proportion of greyish-white pebbles than bones. Whether this is a matter of choice or whether it is dictated by the comparative absence of animal life in its environment is a point which requires elucidation.

Unlike the Satin Bower-bird, the spotted bird sometimes inserts odd decorations among the twigs that form the bower-walls, and most bowers have at least a few shells or pebbles inside them. Often these are sufficiently numerous to form a continuous pathway of decorations right through the bower-passage from one display-ground to the other. It will have been noted that almost all of the articles chosen by the Spotted Bower-bird are to some degree reflective of the harsh inland sunlight. The same is true of the erectile crest of the adults of both sexes.

Following the observation of bower-painting by the Satin-bird, Gaukrodger[89] examined many Spotted Bower-birds' bowers in central Queensland in December, and found one in which the grass stems of the inner walls were coloured reddish-brown for a considerable part of their length. More recently, Chaffer[42] has described painting activities in north-west New South Wales. A bird was photographed nibbling dried grass which, after admixture with saliva, was plastered on the inner walls of the bower by wiping its beak up and down the grass stems, or sometimes by passing the stems between its mandibles. The result was a brown band against the straw-coloured grass of the bower wall. The painted area extended the whole length of the inner walls from about 4 to 8 inches above the ground. Jarman's[133a] Central Australian observations on the western sub-species are of interest. 'Some actions . . . suggested that he was painting the inner walls but, although there appeared to be a deposit of brownish pigment, no definite evidence was obtained.' Painting has been observed only during late November and in December, the period of full spermatogenesis, but it must be emphasized that up to the present these are almost the only months in which it has been carefully looked for. Eight bowers, however, were examined by me in July and August. No trace of paint was found so early in the sexual season.

Display and the sexual cycle

After the hatching of the young in December and January the Spotted Bower-bird deserts its display-ground, and flocks numbering up to 30 or 40 birds forage the countryside in search of fruit. Three crested males were examined from central Queensland flocks in April. The larger testis of each bird measured about $4 \times 2 \cdot 5$ mm. The larger testis of each of two crested males from northern New South Wales during the same month of a different year measured 3×2 mm. In all five organs spermatogonia only were present. Tubules were about $70\,\mu$ in diameter and the tunics were about $80\,\mu$ thick and undistended. Some Leydig cells were already organized into small groups among the comparatively undifferentiated inter-tubular juvenile cells (Pl. 13). The above specimens were taken during periods when nearby display-grounds were not in general use, although the New South Wales collector noticed fresh green berries on nearby bowers.

New bowers have been reported from Western Australia in April,[238] and in Victoria and New South Wales in July[80] and August.[175] It appears that during the period of post-nuptial testis regeneration, Spotted Bowerbirds return to their territories and display there. Gaukrodger, who lived among the birds in western Queensland, believed that the bower was 'always under observation'. We have no knowledge of the events within the flock or of the circumstances under which the female comes to the bower. Robinson says that the western bird erects a new bower every two or three years. This is always built on the former territory, and often within a few yards of the old structure. It would seem that comparatively few of the durable display-things are carried to the new structure by either the western or eastern Spotted Bower-bird.[88, 90, 238] The sites of former bowers can be located by an accumulation of pebbles or bones long after the twigs and grass of the bower have disappeared.

The accounts vary regarding the number of birds which may appear at a single bower. Further, as the period of year at which the observation was made is frequently not given, and the sexes are externally identical (or nearly so), much of the scanty available data is not of great value. Robinson saw as many as six birds watching the display of a solitary western bower-bird at the bower. The same observer records three crested birds together at a bower in April, but they did not carry out sustained display. Robinson asserts that he never saw more than one bird displaying at the same time. There is also a professional collector's account of three or four birds together at a bower in June (mid-winter). Two were fully crested and a third partially so. All were males. Another writer[192] refers to four birds near a display-ground of *C. m. guttata* in June, but no display was observed.

The collector Whitlock,[283] writing of *C. m. guttata* in Central Australia, describes two pairs and a younger unattached male in a bower, but gives no date and no proof of sex identification. The same author has an account of an apparent male displaying to a female and one other spectator in September in inland Western Australia, and later describes a display before an audience of five birds. In this region eggs were laid about six weeks later, and when this occurred interest in the bower waned. Jarman,[133a] too, saw a second uncrested bird perch 'scolding' about 12 yards away from a Central Australian bower at which a crested male displayed to an uncrested bird. Jerrard[134] relates that a third bird visited a Queensland bower where a male displayed to a female. It, too, played a purely passive role. The two passive birds occupied the bower by turns, 'behaving exactly alike and being treated to exactly the same display by their strutting co-partner'. One was crested and the other was not. Here we have a distinct suggestion of polygamy.

Archer[5] asserts that females only are allowed to watch the male display, and that other male intruders are chased away. The collector Jackson, who spent several months among eastern birds, observed displays from the time of his arrival in September, and heightened performances late in October when nests were built. Only once during this period of gonad maturation were more than two birds seen at a bower when, in the first days of October, three of undetermined sex were observed. Here is a further suggestion that the species is polygamous. Other writers,[42, 226] too, have recorded only one or two birds at bowers in November at the height of the sexual season. Gaukrodger declares that he once found two bowers in use about 20 feet apart but he could not state if one male or two owned them.

Most of the brief descriptions of the display at the bower agree, but observers differ widely in their interpretation of its significance. It has been generally suggested that although display and bower-building may have been sexual in origin in the early history of the species, they are carried on today largely for the 'pleasure and interest' that the birds derive from them.[42] The essence of the male display is sudden movement, noise, and the fan-like erection of the coloured nuchal mantle, which gleams silver-lilac as the bird moves in the sunlight. While the male displays energetically—and sometimes violently—with a selected display object in his beak (e.g. a bone), the female watches impassively and usually silently, but nevertheless very intently, from the shadows behind or within the bower. The male displays spasmodically while alone, but it is only when the female is at or near the bower that the display reaches its height of noisy intensity. Jerrard[134] says that if the male attempts to enter the bower the female at once flies out and the male then 'must coax her to return'. Ultimately she flies off altogether. The male, says Jerrard, then stays behind and tidies up the bower in preparation for the next performance. He quietly rearranges the decorations, or refurbishes the walls, or occasionally eats wild fruit that he has brought to the bower. Jerrard believes that the male, rearranging the ornaments, gets pleasure from hearing them 'tinkle against one another'. He will 'take up a shell from the heap and drop it twice or thrice in the same place, listening intently to its fall'.

When the female approaches, the male begins to cavort. He seizes a display object, erects his mantle and, as the performance heightens, raises and lowers the lilac feathers as head and body move in the display. The general plumage is rearranged—it may be fluffed or sleek by turns. The body is stretched, next huddled. The head and wings are thrust into unusual postures (Pl. 11 *f*). The bird frequently leaps high in the air and sometimes 'attacks' violently the display-objects on the ground. While the display proceeds the female steadily regards the male, peering through the interstices of the bower wall if his gyrations shift his position and place it between them. The male contrives to keep his gleaming frill in view of the female and utters 'squeaks, clicking . . . and odd vocal sounds'. Some noises resemble the drawing of corks.[134] The male may pick up a berry, go through the motions of breaking it on the ground, and then lay it on the ground in front of the female before continuing the display. Archer[5] asserts that the male display to the female is accompanied by an 'unending stream of hissing and mimicking'. Chaffer and Ramsay[42] saw a male trip and fall over the pile of bones in his excitement during a display before the female. 'Bower-birds usually hop', wrote Ramsay,[227] 'but when the female arrived the male became so excited that he ran with head up—and fell head over heels.' In the most important observations yet made on the species, Ramsay saw that when the female was present (late in November) the male 'worked himself up into a state of great excitement, attacking the bones and flinging them around'. (Here we see the operation of a displaced combat drive.) One flying bone narrowly missed the watching female, which dodged suddenly to avoid it. The male hopped right over the bower during one prolonged display. After about 20 minutes of sustained display by a male before an intent, though passive, female, copulation occurred outside the bower. This was followed by another violent display during which the male 'aggressively attacked the bones or odd sticks, or leaping high, attacked a tree-trunk'. Copulation occurred again five minutes later, after which the male display continued at a reduced tempo. The female now entered the bower and crouched there motionless for about 10 minutes, and then left the vicinity. When she flew off the male 'began to tidy up the bower'. He did not attempt this while the female was still in the vicinity.

Whitlock's[281] observations at a display-platform mentioned above suggest that the western race performs in essentially the same way, though careful analysis by means of cinematograph films may some day show minor sub-specific differences. Whitlock's western male 'puffed out his feathers on his neck to great advantage. With various harsh cries he advanced into the centre of the cleared space, and made a vigorous attack' on the dried body of a reddish-coloured centipede. 'He advanced and backed, hopped from side to side, pecked vigorously, jumped into the air, and with much apparent ferocity made rushes at one of his immediate audience.' Whitlock took this to be a female. 'Now and again she uttered a short, harsh cry but otherwise seemed to regard the demonstrations made by the male as very matter-of-fact and hardly worthy of notice.' Jarman[133a] saw the display of a western male in Central Australia early in October. The bird 'fanned out his nuchal plume, lowered his head and, with wings half raised and shimmering, moved slowly through the bower. Upon reaching the uncrested female he clicked his bill with hers several times'. Sometimes the male would repeatedly jump into the air with wings outspread, 'almost as though he was jumping against an obstruction'. This latter activity also took place from a low limb. A second uncrested bird perched about 12 yards away, 'scolding repeatedly'. This was the only occasion during 2 days of intensive observation that birds other than the crested displayer were seen near the bower. This bird visited the bower at intervals of ¾ to 1 hour and stayed from 5 to 15 minutes, busying himself in 'straightening the sticks on the inside of the bower and rearranging the decorations'. He always approached by the same route.

Five crested males were examined from central Queensland during September at a period when display was in progress and when fresh green seed-pods were brought to the bowers. The larger testis of each pair measured between 9×8 mm. (two birds) and 9×6 mm. (three birds) in size. All organs presented a similar histological appearance. The tunics were approximately $45\,\mu$ thick and the tubules were expanded to a diameter of $140\,\mu$. The tubules (earliest specimen taken 19 September) all contained bunched sperms but none was free in the lumina. The Leydig cells were still heavily lipoidal but were dispersed so that few remained aggregated. The ovulation date was still probably some 2 months ahead.

The following month four males were examined from south-west Queensland. It is unfortunately not known whether they were crested. During November the bower-display is at its height. The testes of the above birds fell into two groups: (1) Two birds were of an appearance similar to the September material, and (2) the remainder appeared to be approaching the peak of spermatogenesis. One bird, collected on 10 November while renovating its bower, had gonad details as follows: The testes measured 12×7 mm. The tubules had distended to $220\,\mu$ in diameter and the tunica albuginea had correspondingly decreased to a width of $35\,\mu$. The few Leydig cells in each field measured $14 \times 9\,\mu$. Sperms were bunched and some had escaped into the tubal lumina (Pl. 13).

Display occurs spasmodically to a minor degree in trees away from the bower. Two observations by Chaffer reinforce the conclusion that the nuchal frill is epigamic and possibly aposematic in function. Late in November a bird was seen 'eyeing his reflection in a window' of a hut and parading along the sill 'spreading his lilac nape-feathers to the fullest extent', and at times stiffening his body with accompanying vocalization. When a mirror was placed in the bower, the owner, after his first fear, returned and attacked the reflection. When this produced no result he crouched low before the mirror, occasionally picking up a bone and throwing it aside. The nape feathers were partly erected.

Vocal mimicry

Another form of display, that of vocal mimicry of other animals and bush-sounds in general, also occurs in a most highly developed degree. The Spotted Bower-bird is probably the most gifted mocking-bird known. Its original calls are perhaps limited to a ringing, somewhat metallic advertising call, some cat-like cries, and a variety of harsh or hissing notes used in display. However, any deficiency in natural notes is remedied by the remarkably faithful reproduction (to human ears) of those of many other species, as well as the barking of dogs, 'the noise of cattle breaking through scrub', the noise of a maul striking a splitter's wedge,[36] of sheep walking through fallen dead branches, of Emus crashing through twanging fence-wires, and of odd 'camp sounds'.[129] A Spotted Bower-bird mimics a Whistling Eagle so faithfully that a hen and chickens fly for cover, but its mimicry of the Raven was said not to be as hoarse as the original voice.[36] North[205] says that besides bird song, the Spotted Bower-bird imitates the 'whirring-like' noise made by the Crested Bronzewing Pigeon during flight, as well as wood-chopping, the crack of a stockwhip and, in fact, any other sound that it often hears. Most collectors have been so lost in admiration of the bird's powers that they forgot to record when and where the performance took place. Whitlock,[283] however, says that perfect imitations of the notes of other birds are uttered during the bower-display in September. Mimicry during bower-display has been elsewhere recorded;[80] but other accounts[129, 175] make it clear that mimicry takes place also in trees both near and at a considerable distance from the bower. As regards period, it has been recorded at a new bower in Western Australia in April.[238] In eastern Australia it has been described between August (from a vantage point near a new bower)[175] and January (when the birds mimicked in fruit-trees during the post-nuptial flocking period after the bowers were deserted).[129] Striking evidence is available on the employment of mimicry as a distraction-display when the nest with eggs or young are menaced.[89, 129, 281] Gaukrodger says that if an outburst of noisy mimicry failed to engage his attention, the female would 'suddenly almost fall to the ground and with neck stretched out, feathers ruffled and wings spread, she would creep through the grass, pretending helplessness'. Intermittently the mimicry and 'antics of pretence' continued for as long as the human intruder remained still, but if he moved to touch the nest, the bird abandoned her distraction activities and approached closely to hiss, scold, and 'show fight'. McLennan witnessed an excited display of mimicry by a Spotted Bower-bird immediately after he shot its mate in February. The surviving bird flew to a nearby tree and mimicked 'all the birds in the neighbourhood'. Robinson saw a Spotted Bower-bird adopt a threatening attitude and mimic a Butcher-bird and a Whistling Eagle when disturbed at its bower.

The nesting season

We have seen that copulation occurs at the bower in November at the termination of violent display. We do not know what is the female or other activity that causes the sudden transference of the male's physical attention from the bower-decorations to the mate who has for so long waited nearby. It is probable that, as with the Satin Bower-bird, mating is somehow delayed until the environment becomes full of insects of the kind used for feeding the young. The Spotted Bower-bird is largely frugivorous but the young are given a proteinous diet. Gaukrodger[90] says they are fed on grasshoppers, moths, caterpillars, and berries, and there is further evidence that they are at least partly fed on caterpillars.[36] A connexion between the reproduction date and the ripening of the berries of the boobiala tree on which the birds feed extensively has been suggested,[129] but it is more likely that the principal timing factor is the seasonal advent of protein, not special vegetable food. Jackson describes the species capturing swarms of flying insects emerging from under the bark of a coolibah tree in the first days of October, and remarks that the water tank in the collecting area was full of drowned insect life in December, the month of hatching. North has information that breeding is inhibited by excessively dry weather, and Berney[18] reports the complete cessation of breeding of all central Queensland species for 16 months during a prolonged period of drought. Again, Whitlock found that whilst one pair of Spotted Bower-birds in inland Western Australia laid eggs late in October, neighbouring birds left their area and failed to breed when the last water evaporated from the claypan near which their display-ground was built.

There is, then, reason for us to believe that the ovulation date (as distinct from the initiation of the sexual cycle some months before) is governed by the state of the environment, probably by means of stimuli operating through the female exteroceptors which lead to some innate feminine activity that causes the male to shift his physical attention from the display-things to the watching female.'

Differential breeding dates in various localities emphasize this point. In north-west New South Wales and western Queensland the ovulation months seem to be November and December, but no such regularity is evident from collectors' records of breeding dates of *C. m. guttata* of Central Australia, where the sexual cycle of many birds is regulated by spasmodic rainfall or its effects.[135a] Thus, one collector observed young Spotted Bower-birds flying as early as 28 October, and another found a nest with eggs in the first week in February.[205]

The female Spotted Bower-bird builds her nest in a tree or bush within a few hundred yards of the bower. Because of the open nature of the country it is easily found and so collectors' accounts of non-breeding during droughts are probably reliable. The nest is shallow, flimsy, saucer-shaped, and scantily lined with leaves. It is usually between 6 and 50 feet from the ground (Pl. 12). Two, occasionally three, eggs are laid. They have a ground-colour of grey to greenish, and may have darker wavy, thread-like markings, looped and scrolled here and there (Pl. 12). There is a great deal of variation, but in all instances a beneficial disruptive effect is obtained. As the female is apparently not fed by the male, and is compelled to build in open situations, her food-gathering absences leave the nest at the mercy of predators. Hawks and crows abound in the dry regions inhabited by this species.

It is significant that all members of the genus *Chlamydera* that inhabit the scantily timbered, hawk-infested dry areas of Australia are cryptically coloured in the male as well as in the female. The brilliant neck-frill is conspicuous only when it is erected. When it lies flat it has to be looked for carefully or it cannot be seen in the field. The disruptive plumage makes the birds surprisingly difficult to see in the broken light of semi-protected spots where they build their bowers. It is possible, too, that the paler coloration of these inland birds is partly an adaptation to offset the high temperatures (up to 120° F. in the shade) to which they are sometimes exposed.

After the female leaves the bower to build her nest the male displays alone, but the bower is soon neglected, and by the middle of January eastern Australian birds and their young gather into flocks and begin to wander in search of fruit. Gonads from two localities were examined during this post-nuptial communal phase. A single crested male possessed testes which measured 6×4 mm. and which were in a state of tubule metamorphosis corresponding to the refractory period of photo-stimulation experiments. A new tunic had been built up inside the old one and together they were 70μ thick. The collapsed tubules were 90μ in diameter and the new interstitium was in an undifferentiated state with tinges of fresh lipoidal material showing in the cytoplasm of the juvenile cells. Two other specimens were received from north-western New

South Wales. The crest-state was not given, and they presented a condition never before seen in wild birds.*[175]

Discussion

The similarity between the behaviour pattern of the Spotted and Satin Bower-birds is at once evident. Clearly, they are more closely related than is generally believed. Both species, as far as is known, have a prolonged and broadly similar reproductive cycle. Both display during the non-sexual season and take up territory early. The bowers are fundamentally similar. Both choose display-things with great discrimination. The male of each species brings display-things to the bower and postures before the passive watching female whom he in turn watches intently during the performance. In each species only one male displays at a time. Both species paint their bowers. Both display arboreally and mimic vocally. Each uses this latter propensity as part of its distraction display. The principal advertising cry of each is of the same nature. The bower-display of both species is prolonged for weeks or months, after which coition occurs and the young are bred into an environment full of protein food. The nests are very alike, and the egg-pattern is similar although there are more spots than streaks on the eggs of the Satin-bird. In neither species does the male share the domestic tasks of building the nest, incubation, feeding the female, or rearing the young. The male of each species continues his display alone after the female leaves the bower to incubate.[134] The same post-nuptial flocking and fruit-gathering activity occurs in both species.

There are, of course, notable differences between the two species. For example, young Satin-birds are not tolerated at the bowers of fully adult birds after the beginning of the seasonal gonad resurgence. It seems that early in the sexual season the less distinctively plumaged adult male Spotted Bower-bird has no such power of domination over younger birds and unmated neighbours. As spermatogenesis proceeds, however, he appears somehow to become rid of less aggressive rivals. Again, the Spotted Bower-bird adopts a grouping system in his mosaic of bower-decorations whilst the Satin-bird is haphazard in his arrangement, but, on the other hand, the latter species rigorously excludes decorations from within the bower. Further, the mature Satin-bird regularly orientates his bower across the sun's path, whereas *Chlamydera maculata* does not. However, as will be seen in Chapter 6, *C. nuchalis* orientates his bower after the manner of the Satin-bird. There are considerable differences between the two species in the course of actual display, although in both the emphasis is on noise and aggressive activity. In their respective choice of bower-decorations the two species diverge sharply. However,

* A comparable condition has since been found and illustrated in drought-affected non-breeding passerines.[135a]

the dissimilarity between the Spotted Bower-bird's collection of pale pebbles, bleached bones, and shells and other reflecting objects such as glass, metal, and jewellery and the Satin-bird's accumulation of blue, greenish-yellow, brown, and grey display-things is not fundamental. It is a reflection of the materials available in the two environments and of the different colouring of the plumage of the two species. In each case, the male achieves a striking display which is not, however, usually visible at more than a few yards' distance. He provides a conspicuous display for his kind, but not one that will attract his enemies.

We have seen that in the Satin-bird the choice of coloured decorations seems to be dictated by the colours of his mate. The Spotted Bower-bird often uses green berries and cones as do other members of its genus, but these are relatively inconspicuous beside the great accumulation of articles that are pallid or reflective in nature. It is significant that although the coming of Europeans has placed at the birds' disposal an almost unlimited variety of manufactured objects, the bird chooses widely, yet almost limits its selection to metallic objects, and ignores those that are green, the colour of the chosen berries. The crest of each sex is iridescent in the sunshine, and it is possible that the reflecting articles with which the male displays to the female may be chosen because of this. The ventral surface of the female, which faces the male display, is pallid—and so is the bulk of the selected ornaments. At the same time we should not ignore the possibility that the decorations are selected to match the plumage of rival males. As with the Satin-bird, there is a delay of many weeks at the bower before the male's physical attention is finally transferred from the display-things to the watching female.

There is some evidence that display and choice of decorations in the Spotted Bower-bird is innate. Gaukrodger[90] took young from the nest and, at the age of 9 months, they began 'playing' with the vertebrae of sheep, shining tins, and glass-fragments that were supplied to them in captivity. This was in September, a period of high sexual and display activity among the non-captive adults.

It would appear that the development of the similarly cryptic plumage in both sexes is an adaptation to the arid and semi-arid, sparsely timbered environment in which the birds live. The display-ground is usually concealed by low-hanging branches, and the coloured and glistening crest can be seen only at very short range. Even the challenging advertisement call, which resembles that of the boldly plumaged male Satin-bird, is relatively weak. It is not until its eggs or young are menaced that the obliteratively coloured Spotted Bower-bird behaves noisily in the open, and then it is apparently the female alone which does this in the form of a distraction display.

Display and associated phenomena of the Spotted Bower-bird serve the same function as those of the Satin-bird, namely, the attraction and retention of a female, the defence of the territory, and the development and synchronization of the joint sexual processes until the environment reaches a seasonal state suitable for successful reproduction. As with the Satin-bird, there is available no direct evidence of polygamy.

It is not of course suggested that *Ptilonorhynchus* and *Chlamydera* are anything but true genera. But it is strongly held that despite opinions that 'bower-building habits seem to be imitative and adaptive, and [apparently] do not indicate close relationship',[185] the common ancestry of *Ptilonorhynchus* and *Chlamydera* is perfectly obvious. Further, as will be shown in Chapter 22, the radiation of *Chlamydera* has occurred in comparatively recent times.

6

GREAT GREY BOWER-BIRD
Chlamydera nuchalis (Jardine and Selby)

THE Great Grey Bower-bird is the biggest of the family. It measures between a little under 14 inches long in north-east Queensland and up to almost 15 inches in north-west Australia.[191] It lives in an enormous sweep of humid and sub-humid open tropical scrub-land which has an annual 'wet-season' rainfall varying from about 20 to 60 inches. The bird's range (Fig. 7) includes much of the country northwards from a rough line extending from the Burdekin Valley (lat. 20° 10′ S.) on the Queensland coast right across the continent to the Fitzroy River (18° 10′ S.). The Great Grey occurs above the arid western desert behind the Eighty Mile Beach; the Spotted Bower-bird lives in the country below it. On the Queensland side, the Great Grey inhabits the valley of the Burdekin. It is once more replaced by the Spotted species in the valley of the Isaac River, immediately south. This fact was first recorded in 1845 by the explorer John Gilbert who was speared by aborigines on the Mitchell River later in the same year.[48]

Across the long dry lead from the Burdekin to the Fitzroy the precise partings of the ways of the Great Grey and Spotted birds are unknown. It is probable that for many miles the range of the Great Grey approximately follows the southern boundary of the wide sub-humid scrubby belt that runs north of the Selwyn Range. However, the Spotted Bowerbird occurs on the Leichhardt and Cloncurry Rivers north of the Selwyn, so both species live on the broad watershed from which rise the numerous rivers that flow northward into the Gulf of Carpentaria. The Spotted bird can exist in country with an irregular rainfall of not much more than 5 inches per annum. It occurs, in Central Australia, as far north as the Macdonnell Range. In this inhospitable region the Great Grey Bower-bird is unknown. Its adaptation to the periodically saturated, open, scrub-lands has been so successful that if it breeds in a locality it is generally not long before birds, or a bower, are seen. In the extreme north-east of Cape York the Great Grey Bower-bird is replaced by the rarer Fawn-breasted Bower-bird (*C. cerviniventris*) which inhabits also the isles of Torres Strait, the Louisiade Archipelago, and the coastline of New Guinea.

Mayr and Jennings[191] recognize races of *C. nuchalis* from Port

Dennison (*C. n. orientalis*), Port Darwin (*C. n. nuchalis*), Kimberley (*C. n. oweni*), and Cape York (*C. n. yorki*). Although speciation has proceeded to some degree, none of the above, of course, has separated to anything like the extent exhibited by the eastern and western populations of the Spotted Bower-bird.

As with other early discovered bower-birds, the type locality of the Great Grey bird is uncertain. The pioneer Sydney naturalist Macleay accumulated a series of bird skins of which several important ones were inadequately labelled. Among these was the first specimen of the present bird. The skin was sent to the Linnean Society of London. It was described and named *C. nuchalis* by the Englishmen Jardine and Selby in the year 1830.[132] Gould believed that the specimen was a West Australian bird, but whether he had any proof of this, or said so because similar specimens were brought from Western Australia by officers of H.M.S. *Beagle* (in which Charles Darwin served), has never been ascertained. The first description of its bower (Pl. 14) was probably furnished by Captain Stokes,[265] who on the Victoria River at first thought it was 'some Australian mother's toy to amuse her child'. Later, at Port Essington he was invited to see a bird's 'play-house'. He recognized the structure on which he had 'found matter for conjecture' some time before. He now 'found the bird amusing itself by flying backwards and forwards, alternatively taking a shell from each side and carrying it through the archway in its mouth'.

The adult male Great Grey Bower-bird is brownish above, with ash-grey feather-tips that give its back and wings a mottled appearance. It has grey underparts. The male has a slightly dark-mottled crown and an erectile lilac crest. This crest is similar in colour and microscopic structure to that of the Spotted Bower-bird. In this connexion a field observation of Sedgwick[251] is of interest. 'The bower-birds' nuchal feathers were erected. . . . These feathers appeared orange against the light, but each time the bird moved into the shade their true lilac coloration became apparent.' Actually, as I have shown, the 'true' coloration is orange.[175] The lilac, and next the silver, is a physical phenomenon.

The female is slightly smaller. She is paler above than the male, and lacks the highly contrasted feather mottling. Below, the sexes are externally almost identical. The adult female has a slight dark marking on the forehead, and often lacks the nuchal crest completely. Young birds of both sexes basically resemble the adult female, although there are differences between them. It is not known at what age the young develop their crest.

The Great Grey Bower-bird feeds largely on wild figs, wild 'plum', and other fruits; also on *Acacia* and other seeds. There is plenty of information about the variety of fruit—paw-paws, granadillas, guavas, mangoes, peas, tomatoes, chillies—that they take when they visit homestead and suburban gardens in tropical Australia. I have found Coleoptera and Hymenoptera in a stomach. There is a description of a Great Bower-bird taking a hen's egg,[57] and another of five birds which succeeded in getting past the wire screen in a cattle-station butcher's shop. They were found feeding on fragments of meat left on the chopping-block.[284] The species is never found far from water.

The bower and the display-ground

The display arrangements of the Great Grey Bower-bird are on much the same plan as in the Spotted species. The bower, like its owner, is bigger (Pl. 14). It is built on a low flat platform of twigs. It is commonly from 2 to 3 feet long and is built of interwoven coarse dry grasses and twigs or sometimes of twigs alone. The walls are about 4 inches thick. They are generally about 18 inches high and arch inwards, often meeting over the central avenue which is some 6 inches wide. There is a reference to a bower of some kind being built on the roof of a veranda.[57] This recalls the description of a platform built by the Spotted Bower-bird on a broad tree-limb in Victoria.[80]

At each end of the bower is an extensive display-ground which is covered by a variable quantity of sun-bleached or other generally pale objects. The bulk of these are kangaroo or wallaby bones, land-shells, or pebbles of quartz, limestone, or laterite, but if the bower is within a few miles of the sea, marine shells may be the principal decoration along with stray water-worn fragments of calcareous corals and Polyzoa. Sedgwick estimated that at least 800 land-shells (*Thersites* sp.) occurred on a North Australian bower. This number, he believed, involved assiduous collecting over a wide area. After 453 shells were counted at one end of the bower they were left in small groups of ten. The following week the shells were found regrouped at a new bower built 5 feet away. Silvery leaves and white parchment-like cocoons are often used. The relative proportions of different kinds of display-things depend a great deal on their availability within individual areas. In the roaring mining days of the last century quantities of pale glass were available, and were used by the birds together with other highly reflecting articles such as nails, brass cartridges, tea-spoons, occasional sixpenny pieces, tin tobacco tags (stuck into hard-packed plug tobacco favoured by old-time bushmen), metal buttons, knife-blades, and so on. In the Ord River country we are told that 'Mr. Edward Delaney missed his spectacles'. They were eventually found near a display-ground.[136] Scraps of white paper are taken to display-grounds. Whitlock[284] noticed that many pieces of light green and clear glass were used, but that nearby bottle-green and brown glass was ignored.

Some miners did not like the birds 'as they pilfer any small bright articles lying about the camp'; also for the depredations they committed in their gardens.[205] Other miners and charcoal-burners made pets of them, and the birds 'freely entered the tents'.[66] There is also a more recent record of a Great Grey Bower-bird entering an army supply hut and stealing a large number of roofing nails which it took to its display-ground. Broken insulator caps are collected by the birds. Day,[66] a veteran collector, found 'a bright specimen of gold' embedded in one glistening piece of quartz, and recalls that in opal-bearing country he used frequently to find pieces of precious opal at the display-grounds. On the Leichhardt River, after making a meal from wild pigeons, Day saw Great Bower-birds snatch up the bones as soon as he discarded them. When built near the remote towns of northern Australia, the bowers may contain an almost inconceivable accumulation of pale or reflective rubbish. Yet this is chosen with great discrimination. The military occupation of the north during the last war provided the birds with an unprecedented harvest of treasure. Brass cartridges of various kinds, metal identification disks, machine-gun links, aluminium, tinfoil, paste-tubes, and so on were taken to bowers.

A few green objects are regularly chosen but not green manufactured articles. Thus, most observers have found fresh green leaves, green cones, small green fruit, sometimes pale green flowers, green plant galls, but not green cloth, paper, or painted metal of similar tones. Hopkins[118] says that wild passion-fruit are used when young and green, but not after the first yellow tinge appears. Actually green seems to be the only pure colour consistently used, though there is one hearsay account of Townsville bower-birds 'snapping off and carrying to the bower rare, lilac-coloured blooms' which they stole from an orchid fancier.[117] There are one or two vague collectors' stories that coloured stones or feathers have been observed but the colours were not stated. On Cape York I found that the birds decorated their display-grounds with grey laterite, white quartz, greyish cocoons, and green cones, but never used any of the profusion of golden orchids which clung to the *Melaleuca* trees all around. Various people have put coloured objects on bowers, but these were removed by the owners. Such included reddish plastic identification-disks, and rag or paper of different hues. Hopkins,[118] however, describes a particular display-ground which, built near human habitation, was decorated with wool, glass, plastic cord, scraps of cloth, a plastic bottle top, and a narrow leather belt—all red in coloration.

Although the bird removes all fallen leaves from the display-area, it will allow green cones and fruit to wither and discolour without discarding them. A zonation of decorations has been observed. Thus, land-shells may be grouped together, and placed apart from the kangaroo bones.[251] Many bowers have decorations within the central avenue and some have them strewn right through from one end to the other.

In one striking respect the Great Grey Bower-bird differs from all other members of its genus—it consistently leaves standing the bowers of former seasons. As many as nine[120] structures have been found under one tree, though this is quite exceptional.

After examining a series of such groups I had the impression that the bird usually builds a new bower every year. Of four groups inspected on Cape York, three contained three bowers and the fourth was composed of two. In each case the relative ages of the bowers were easily judged. The month was September, and fresh twigs and green cones made it certain that one of each group of bowers was in use, even though no bird approached during the brief periods of examination. In each assembly there was a new bower (with fresh green display-things), as well as an older one (still in good condition and often containing last season's collection of quartz and laterite). Some groups also contained a very old bower on which usually remained no decorations. The twigs of this bower were snapped off cleanly near the level of the platform. Presumably these had been used in the construction of newer bowers. It will be recalled that the Satin Bower-bird has this habit. There was no evidence that any but the newest bower of each assembly was currently in use for display.

Altogether fourteen bowers or their remains were examined and it was found that of those whose orientation could be unquestionably determined, all (no less than thirteen) were pointed within 45° of due north and south. The mean deviation was only 16°, and in the grouped bowers there was a most striking tendency for all within a single assembly to be orientated in a similar way. Thus, one group of three bowers showed a deviation of only 13°. A group of two were pointed within 1° of each other. Another assembly of three contained two bowers on a bearing of due north, and the third on 350°. The overall orientation was so striking that full details are included below.*

It is not known whether such fidelity to a north–south orientation is widespread elsewhere. Sedgwick, writing of North Australian bowers, mentions that a new bower was 'erected two and one-half feet away in the same direction as the earlier sites'. Selvage[252] took bearings on five Townsville bowers of which 'four had the passage running North and South, the other East and West'.

The Cape York bowers described above were built several—usually about 5—yards apart in semi-shade in open *Melaleuca* or other scrub. Sometimes they could be seen at a distance of about 60 yards. Whenever

* In a brief war-time note[163] a summary of these data was wrongly given as referring to *C. cerviniventris*.

possible, however, the Great Grey Bower-bird builds its bower in a well-concealed position. In the Larrimah-Katherine area Sedgwick found bowers built within a few feet of each other in ebony thickets. Here the frequent clearing of sections of thicket for the accommodation of troops provided the opportunity for brief but significant observations concerning the territorial attachment of the species. When a bower was interfered with, the birds did not desert the area but either retained the bower *in situ* or moved it to the nearest available cover. On one occasion, when

Bower group	Bower number		Bower bearing
(a)	1	(single bower)	360°
(b)	2	,, ,,	6°
(c)	3	,, ,,	45°
(d)	4	new bower	33°
	5	older bower	10°
	6	oldest bower	20°
(e)	7	new bower	35°
	8	older bower	34°
	9	oldest bower	*
(f)	10	new bower	350°
	11	older bower	355°
(g)	12	new bower	360°
	13	older bower	360°
	14	oldest bower	350°

* Bower too old for an unquestionably accurate bearing to be taken.

all cover was removed from a 'fairly considerable area', the birds remained in their territory and built in the open. When a small clearing was made to within 3 feet of a bower, a new one was built in a small thicket 35 yards from the original site. The display-things were also transferred. A bower was built under a wooden, plank-covered staging in what had been a large camp, but in which only a few troops then remained. Later a larger body of troops moved in, and the staging was removed. The bower was then rebuilt about 10 yards away, where 'branches of a small tree gave token shelter only'. The Great Grey Bower-bird's desire for bower-cover, when such is available, sometimes leads it to build inside deserted human habitations. The pioneer collector Keartland[135] found a bower and display-ground built in an old native *wurli* or hut, and Condon[53] found a flourishing display-ground in a deserted army hut.

Display and its season

It is not known at what period the annual sexual resurgence begins. On the meagre data at our disposal it is impossible to determine which of the recorded bower-constructions and displays refer to irregular post-nuptial activities, and which occurred after the onset of the sexual season proper as indicated by the gametogenetic modification of the primary organs of reproduction.

Gonads have been examined from birds taken in periods of both quiescence and bower-building. Testes from an uncrested flocking male collected by me in the Gulf of Carpentaria in June 1942 measured $3 \times 1 \cdot 5$ mm. A non-displaying crested bird, collected by J. A. Keast on the Forrest River, Western Australia, on 30 May 1952, had testes of similar size. The seminiferous tubules in each bird measured only $50\,\mu$ in diameter. They contained spermatogonia, but in the Forrest River bird a few primary spermatocytes were also present. The testis tunics measured about $120\,\mu$ thick. In the Forrest River bird, numerous Leydig cells had become embedded in the tunic during its post-nuptial rehabilitation of the previous season. In each bird the Leydig cells measured only $5\,\mu$ in diameter. The first-mentioned gonads were fixed in a flask of whisky and the testis lipids were consequently lost during the alcohol grading of wax sections, but the formal-calcium fixation of the Forrest River specimens permitted the examination of a heavy deposition of cholesterol-positive lipid in the Leydig cells (Pl. 13). Further, the seminiferous tubules were already clear of fatty post-nuptial debris. This, and the presence of primary spermatocytes, seemed to suggest that the new season's sexual resurgence had just begun.

Three crested males were collected at or near newly decorated bowers by G. Gibbs in the Townsville-Charters Towers in September–October 1952. All measured from between 10×7 to 13×9 mm. and contained bunched spermatozoa in tubules from 200 to $220\,\mu$ wide. Tubule expansion had reduced testis tunics to a thickness of $20\,\mu$ or a little less. Leydig cells were considerably dispersed (Pl. 13) and individually measured from 10 to $12\,\mu$ in diameter. In no case had sperms been shed, nor were quantities of them yet free in the tubule lumina. We have evidence, therefore, that the display of yet another bower-bird is probably related to the reproduction cycle.

New bowers have been observed in February (south of Darwin),[53] and, after disturbance or destruction of the former bower, early in March and in 'April or May' in the same general area.[251] Unmistakable bower-display has been witnessed in June and display at the bower by a bird with an erected crest in August.[251] Selvage[252] says that near Townsville a bird built its bower in a school playground every year in February.

The display of the Great Grey Bower-bird has never been fully described because most of those with opportunities to do so were interested only in the collection of skins or egg-shells, or were engaged in duties that afforded little occasion for non-military diversions. Hopkins[119] says that the female stands either near, or on a branch above, the display-ground, whilst the male displays by 'rushing through the bower, or strutting around with wings slightly drooped'. Sedgwick says that the bird utters a harsh purring sound, varied by a low cry, while renovating its bower (late in June). Increased bower-display was accompanied by a loud 'ticking' sound which is uttered with a bobbing of the head. Another form of display involves a stiff, hopping movement with wings drooped and head depressed. On another occasion (early in August) three or four birds were in the vicinity of a bower but no more than two were actually at it at any time. 'From time to time a bird would pick up an ornament or a stick ... and run around or through the bower. Occasionally a bird so engaged would break off to indulge in what was, apparently, a friendly chase.' The participating birds hopped over the ground with tails slightly raised. 'The clicking or ticking call ... accompanied the display, and nuchal feathers were raised when excitement became intense.' In April a large number of birds was present, and one male spread out its crest 'until the feathers resembled a widely extended fan, almost forming a circle'. Rogers says that most display takes place early in the morning and again just before sunset; the birds avoid the heat of tropical mid-day. Each morning the leaves that had fallen during the night were removed. The leaves were carried away with a 'peculiar sidling dance motion'. An undated observation by the same collector suggests that the male 'plays for hours ... uttering strange noises'. One such call resembles 'a piece of silk being crumpled or shaken'.[240] The female was described as undemonstrative, preferring 'to hop quietly about or sit in a bush'.

Mimicry has been heard, but little of the data indicates the times and situations in which it occurs. One informant[252] describes a Great Grey Bower-bird mewing like a cat from a post under which a cat slept. 'It kept this up until the cat got up, glared at the bird and stalked off.' Another bird whistled like a hawk, and a flock of doves that were feeding near the bower 'dashed into the sheltering branches of a shrub'. Another writer says the birds give only 'passable imitations' of the calls of other species, and that some of their notes are ventriloquial. There is another account[50] of birds that frequented a garden near Charters Towers and learned to imitate the sound made by a lawn-mower, and the clashing, rustling noises made by the homestead deer when fighting.

The nesting season

We have no knowledge of the factors that lead to mating, or what, if any, part the male plays in subsequent events. The nest is a wide (10 × 10 inches) saucer-shaped structure, loosely built of slender twigs. Its egg-cavity is about 6 inches in diameter and $2\frac{1}{2}$ inches deep. The nest is lined with a few leaves or fine twigs but the eggs can often be seen from below. Possibly because of the lack of tall timber, it is usually built between 1 and 15 feet from the ground.[241] It is generally found easily because of its bulk and the open nature of the vegetation. One egg is perhaps more usual, but two are often laid. These are oval and of a pale greyish-green ground-colour, with hair-like markings of brown, blackish-brown, umber, and purplish-slate colour which twist in all directions and often encircle the egg. When a nest is destroyed the female does not appear to attempt to breed again that season.[136]

The bulk of egg-laying occurs from October to December inclusive, but there are interesting departures from normality. An egg was seen in the Fitzroy River region in July,[258] and in North Australia ovulation has been recorded in August.[211] It is said that September layings are not very unusual. Olive[211] has suggested that the breeding season is governed by the seasonal rains, and stresses the fact that the August ovulations mentioned above were in 1895 when unseasonal rain fell during that month. Another veteran collector, Keartland,[135] asserts that rainfall influences the breeding time from September to December in north-west Australia. And, in fact, almost all egg-dates do fall within the general period when torrential rains sweep across northern Australia.

A post-nuptial flocking phase occurs. I saw a flock of nine in June. The old-time collector Broadbent[46] records having seen between 20 and 30 birds feeding together. He gave no date. The collector Rogers,[240] working on the Fitzroy River at the end of the last century, said that the birds left the area late in February, and early in March the bowers fell into disrepair. After a few days' neglect the bowers appeared as though they had been deserted for months. On 8 April a large number were found at a bower and one bird displayed its 'ruff'. Moult, incidentally, has barely begun in January.[240]

Discussion

Although the Great Grey Bower-bird has a geographical range far less extensive than the Spotted species, it is the more successful bird of the two on the simple criterion of relative abundance. This is probably because it inhabits areas which, however dry and desolate they may be for a great part of the year, are annually assured of torrential precipitation and a consequent superabundance of the protein food that is no doubt necessary for the rearing of the young.

Collectors have stated that the rainy season governs the breeding season. This is undoubtedly true in that the species hatches its young into a wet environment that is teeming with easily captured insect food and, if the young were launched at any other time, the species would become extinct. But there is at present no evidence that it is rainfall *per se* that causes the initiation of gametogenesis. Neither is there any proof that actual precipitation stimulates mating. The species is common at

Townsville, north Queensland, where the average monthly rainfall in inches is as follows:

TABLE III

Aug.	Sept.	Oct.	Nov.	Dec.	Jan.	Feb.	Mar.	Apr.	May	June	July
0·50	0·71	1·18	1·95	5·02	10·91	11·83	7·87	3·24	1·20	1·32	0·72

(Calculated over a period of 80 years from 1871 to 1950.)

These figures do not, of course, apply to the whole wide range of the Great Grey Bower-bird, but they are applicable as a general guide to the time when the monsoon sweeps across tropical Australia. We have seen that ovulation generally occurs from October to December, although eggs are not infrequently laid in September. The scanty evidence available from Townsville suggests that nest-building occurs there in October. Because the seasonal rise in rainfall generally begins in September, and more than doubles the August precipitation in October, it could be suggested that this increase, in a very dry environment, is probably sufficient to stimulate the birds towards the end processes of display and perhaps to nidification and ovulation. This matter will be unquestionably proved only by experimentation and, in addition, meticulous observation of nesting and actual ovulation dates in relation to natural fluctuations in rainfall. In 1947, for example, a rainless July was followed by 3·25 inches in August. Theoretically, reproduction should have been early—but no data are available. In 1944, September, October, and November provided a total of less than 1 inch of rain. December brought in a deluge of almost 9 inches. Thus, in that year breeding should have been late—but again no data are available. As Great Grey Bower-birds occasionally build their display-grounds in the private gardens of the residents of Townsville, it is perhaps not too much to expect that careful observations may some day be made.

The rainy season ends in March or April. There is apparently sufficient fruit and occasional protein-food available during the dry season to support the birds even in the most barren parts of their range. If we look casually at the environment it seems remarkable that the species can habitually lay only one or two eggs each year and yet flourish. Its cryptic coloration no doubt protects it against raptorial birds. Although goannas, and probably other reptiles, almost certainly eat the eggs and young it would seem that the Great Grey Bower-bird can have few other dangerous enemies. Perhaps a principal reason for its prevalence is the aborigines' highly civilized attitude towards it. The eggs of the bird are often at hand's reach from the ground, and the adults are highly vulnerable while engrossed at their bower. Yet the native people do not molest them. 'The bowers of these birds are often built close to native camps', Thomson[273] has written. 'The aboriginal does not destroy life wantonly.'

It is regrettable that the same cannot be said for the white savages who destroy the Spotted Bower-bird in eastern Australia because it is a minor pest in homestead gardens. The stupid slaughter of about eighty in one week has been reported from one small township alone.[76]

We have seen that the Great Grey and Spotted Bower-birds are very alike in so far as the habits of the former are known. The absence of a nuchal crest in a proportion of the females of the Great Grey is a matter of profound evolutionary interest. Both species are clearly derived from a common stock. It seems easier to believe that the Great Grey females are in the process of losing the crest rather than gaining independently what appears to be an almost precisely similar, though extremely complex, adornment. The factors that control the colour-choice of the present species are probably similar to those operating in the Spotted Bower-bird.

7

FAWN-BREASTED BOWER-BIRD

Chlamydera cerviniventris, Gould

IN Australia the Fawn-breasted Bower-bird has a restricted range in one of the remotest parts of the continent and, because of this, it was the last mainland member of the genus to be discovered. The first specimen was a male shot in its bower by John MacGillivray,[155] naturalist to the survey ship *Rattlesnake*, on Cape York in October 1849. *Rattlesnake*, in which the young T. H. Huxley sailed as surgeon, collected also a shipwrecked Scotswoman who was living with the aborigines in the same area. MacGillivray carefully described the Scotswoman, but not the new bower-bird; a thing I cannot understand. He had been told that some bower-birds were seen in a 'thicket or patch of low scrub' half a mile from the beach and after a long search he found a new bower, 'four feet long and 18 inches high, with some fresh berries lying upon it'. It was built near the border of the thicket where the bushes 'were seldom more than ten feet high, growing in smooth sandy soil without grass'.

Next morning MacGillivray was landed before daybreak, taking with him a native and a large plank on which to carry back the bower. He was compelled to load one barrel with ball in order partly to relieve the anxiety of the black boy, who feared the threatened arrival of a war-party from a neighbouring tribe. The two men crouched in the scrub within shot of the bower, and caught several glimpses of a shy greyish bird. 'It darted through the bushes in the neighbourhood of the bower, announcing its presence by an occasional loud *churr-r-r*, and imitating the notes of various other birds, especially the Leatherhead.' MacGillivray said that he had never met a more wary bird. For a long time it enticed him to follow it a short distance and then flew back 'to alight on the bower. It would deposit a berry or two, run through, and be off again (as the black told me) before I could reach the spot'. At length MacGillivray saw the bird enter the bower once more and, within range, fired at the bower itself. He killed the bird. When *Rattlesnake* reached civilization MacGillivray wrote to John Gould: 'You will oblige me by comparing the *Chlamydera* from Cape York with the other members of the genus, as I have a strong suspicion that it may be different.' It was different, and at the 1850 meeting of the British Association Gould described the new species as *C. cerviniventris*.[103] MacGillivray's original bower went to the British Museum.

The Fawn-breasted Bower-bird is one of the few continental bower-birds about which there is any precise knowledge of the original discovery. Many years later identical birds were found in coastal New Guinea, in the islands of Torres Strait and, it was reported, in the Louisiade Archipelago.[70]

The bird is about 11½ inches long. Neither sex has the lilac nape-frill found in the other Australian members of *Chlamydera*. The sexes are externally almost, if not quite, identical. Dorsally the bird is dark ashy-brown. Its ventral surface is fawn, with inconspicuous darker markings. In Australia the species occupies only a small strip of coastal country extending from the Lockhardt River area (lat. 13° 04′ S.)[273] on the eastern coast of Cape York Peninsula to the tip of the Cape some 150 miles northward (Fig. 7). This is a low-lying sandy land, flanked by mosquito-ridden tidal mangrove swamps, and seasonally saturated by about 60 inches of rain. It is clothed principally with stunted open *Melaleuca* and *Eucalyptus* forest and scrub, and bisected by streams bordered by a luxuriant tangle of pseudo-jungle composed of vines and other vegetation of northern affinity. The precise extra-continental range is unknown. In southern New Guinea (Pl. 15) its distribution follows the coastal country from at least as far westward as the Oriomo River[188] (west of the Fly) right along to Milne Bay in the south-east, and up the northern coast to at least as far as Humboldt Bay and Lake Sentani. In New Guinea it commonly inhabits timbered grass-lands up to 1,500 feet and it has been found higher than 4,000 feet.[229]

According to the Jardines,[133] pioneers of Cape York Peninsula, the Fawn-breasted Bower-bird feeds largely on a small, reddish-black fruit known locally as the 'native-grape', which grows in the vine-scrub. Berries have been found in stomachs in both countries. Everywhere the birds are evenly distributed. In Australia they seem nowhere very plentiful, but in New Guinea they are reported to be common in certain areas.

The bower and display-ground

Most bowers are said to measure about 14 inches long. The walls are 12 to 15 inches high, 4 to 6 inches thick, and are 'inclined to arch'. The central avenue is from 3 to 5 inches wide. There have been special comments concerning the 'weaving' of the inner parts of the walls. So closely are the sticks and very fine inner twigs interwoven that the walls of the avenue are unusually smooth. The bower, as is usual with the avenue-builders, is wedged into a platform. The display-grounds at each end measure from 3 feet to 3 yards in diameter and are often unequal in extent. Here the Fawn-breasted Bower-bird sometimes differs radically from its mainland congeners—its platforms seem to be generally much more massive and, as a consequence, raise the bower clear above the ground. Cape York display-grounds are sometimes only 2 or 3 inches high, but those in New Guinea may be as high as 14 inches[286] (Pl. 15). It seems probable that the same platform is used for years in succession. Such elevated structures may have value in protecting the bower and display-objects from the torrential rains that cause periodical flooding of northern lowlands.

The crestless Fawn-breasted Bower-bird shows a most intriguing departure from the crested *Chlamydera* stock in its choice of bower decorations. We have become familiar with the large accumulation of bleached bones, shells, and grey and white fragments of quartz, laterite, limestone, and reflecting pieces of glass, as well as other man-made objects gathered by other members of the genus. On the bowers of the Fawn-breasted birds bleached shells, bones, and fragments of the carapaces of crustacea have been found, but these are never present in anything like the quantities employed by the two crested species. Most bowers lack them altogether. The principal display is always a collection of green berries. The younger MacGillivray,[185] who examined seven bowers at Lloyd Bay, said that 'the only decorations about these bowers were bunches of green berries—no shells, flowers, or leaves were noted'. Glossy green berries were found on every bower. They were placed on either the platforms or in the avenue, or stuck on the twigs along the tops of the side walls. One hundred berries were counted on a single bower. The old and withered berries were removed to a heap outside the display-ground. One bower was built in the mangroves below the high-water mark, and was flooded by the tide every fortnight. There was a shelly beach 3 yards away but shells were never used—'always the same sort of berry'. Other writers, too,[16, 280] report green berries, either singly or in small bunches, as the principal decoration. Thomson,[273] who extended the known range of the species to Bare Hill, south of Lloyd Bay, commented on the raised platform. He stated that the Fawn-breasted Bowerbird's display-ground is not as elaborately furnished as that of *C. nuchalis*, a species with which he was even more familiar. He found that the Fawn-breasted birds' bowers were 'adorned chiefly with green fruits broken into fragments, leaves of a silver-grey colour, and the immature flower shoots of a *Melaleuca*, with a few pieces of land mollusc shells. . . .' In New Guinea a bower was described[286] with uniformly green berries and leaves as decorations. Crandall[59] found a New Guinea bower in the rain-forest. Its entrance was covered thickly with 'round green seed-pods about the size of buckshot. Over these were scattered larger pods of two different types, and a green fruit very like a gooseberry [Pl. 15]. Several sprays of the smaller sorts were hung on the twigs of the bower, including the passages. On the foreground, a short distance away, was a rubbish heap formed of a considerable quantity of fruits and seed pods, in various stages of decay.'

Rand,[229] who collected a series of skins in New Guinea, says that although the birds frequent savannah country, the bowers are always concealed by vegetation. Where the lightly timbered grass-lands are contiguous with rain-forest the bower is sometimes built just inside the heavier timber. One of the display-platforms, says Rand, is usually larger than the other, and contains most of the decorations. A few decorations are placed in the walls, some in the avenue, and a few on the smaller platform. The ornaments are 'usually green fruits; occasionally a few green fleshy leaves were also used'. Withered fruit was replaced with fresh.

Rand saw only one bird at the bower. He placed a stuffed female skin at one end of the avenue. 'Upon the male's return he at once attempted copulation' with the stuffed bird. This occurred again after about ten minutes. After this the male spent some hours merely watching the substitute from a little distance away.

No information is available concerning the initiation or duration of the display season. Rand took three birds 'in breeding condition' in September and one in December. Apart from his notes above, little is known about what may take place at the bower. The bird is said to be 'very noisy at its playground'.[273] Jardine says that it displays just after daybreak and again at about sunset, when the birds run through and around the bower with their primaries trailing along the ground. They stop now and then to pick up a bone, feather [*sic*], shell, or berry while their companions are perched in the neighbouring trees 'uttering all the time their peculiar notes'. MacGillivray (the younger) said that while displaying, the bird made a 'rustling noise with its throat'.

As mentioned by its discoverer, this bower-bird, too, is a mimic. One writer[286] says that it mimics while displaying at the bower, and claims that it rivals the Lyrebird as a mocking-bird. Its song in general has been called 'weak but penetrating, and consists of a variable number of harsh grasshopper-like notes, the whole being overlaid by a kind of "hoarseness". The song is slightly ventriloquial in effect, and appears to originate nearer the observer than is the bird.' The bird uses a harsh scolding alarm call.[275]

The nesting season

Ovulation on Cape York has been recorded in November,[205] and a fully fledged young bird was found in the middle of January. The nest is a bulky open cup-shaped structure composed of thin sticks, twigs, and bark. It is 8 to 10 inches in diameter and has an egg-cavity 5 or 6 inches wide and 2 or 3 inches deep. It is usually placed less than 30 feet high in a tree or pandanus palm. A single egg is usually laid by Cape York birds, but two have been recorded in New Guinea. The eggs have a pale creamy-white ground-colour, and are prominently marked with a labyrinth of lines and hair-like figures that twist and zigzag in all directions, many completely encircling the egg. These are umber, olive-brown, blackish-brown, and purplish-slate in coloration. The eggs are swollen oval in shape. In New Guinea Rand found a nest containing young in December. It was about 150 yards from a bower but it was not discovered whether nest and bower belonged to the same pair of birds.

There is some evidence that a minor post-nuptial flocking occurs. The few people who have seen the birds say that they are seen mostly in pairs, or family groups. However, small companies from 4 to 8 in number are sometimes seen traversing open parts of the forest in search of food. There is a record of a new bower being built in January, 17 days after the former one was removed for transport to a museum. It was built a few feet from the old site.

Discussion

The Fawn-breasted Bower-bird is of outstanding interest. Although a crest is lacking, no responsible person has ever questioned that it is a true *Chlamydera*. Apart from its physical similarity to others of the genus, it builds much the same sort of bower (though often raised higher from the ground) as the other Australian members. It builds a similar nest and lays generally similar eggs. It is an accomplished mimic. As far as can be ascertained from the available data, it eats the same sort of food as the other species and should be a direct competitor wherever its range overlaps with that of *C. nuchalis*. It is not uncommon in the narrow strip of low-lying tropical coastline to which it is restricted. It has become 'very common' in parts of New Guinea, where *C. nuchalis* does not occur. It remains successful, then, in its present environment. In fact, unless the Cape York record[185] of bower-building in a regularly flooded mangrove swamp is exceptional, it would seem that population pressure is driving the bird to expand its range in any way that it can.

It is interesting that, like *C. nuchalis*, the Fawn-breasted Bower-bird lays only one or two eggs. It would seem that the usually single annual ovulation is not here an adaptation to the food supply for the young. The bird brings out its young during the wet-season when the environment is teeming with easily caught insect food. Further, hatching occurs in mid-'summer', when the time allowed for food-gathering is greater than at any other period. It is reasonable to assume that there is no lack of day-length to prevent the capture of sufficient food to sustain a greater number of offspring. Thus we may have to look for some totally unsuspected influence to account for the low clutch size. It is possible that in the dry non-breeding season on the mainland both Fawn-breasted and Great Bower-birds may not easily find sufficient food to keep a large population alive and that, in the absence of predators, some kind of selection operates at the species level, and has led to the evolution, and retention, of a small clutch. It would seem that predator pressure must be low, and individuals long-lived, or else more eggs would be necessary to keep the species in existence.

Another fascinating question is this crestless bird's partial, and often complete, abandonment of bleached and other pale or reflective objects as bower-decorations. We have seen that the two crested species avidly collect white and especially reflective display-things. Berries, a type of ornament found on the bowers of all true bower-birds in both countries, are used very sparingly indeed. On Cape York Peninsula, bleached shells are as common as anywhere else, and bones, quartz pebbles, and other white objects occur in plenty. Thus it would seem that lacking the pale chest and the crest (which glistens silver in the sunshine), *C. cerviniventris* lacks also the insatiable drive to collect pale, reflective bower-decorations. It would be easy to settle this point experimentally.

Rand's experiment with the mounted skin appears to indicate that, despite reports that the bower is primarily a communal 'playing-place', it has, in fact, the same function as those of the Satin and Spotted Bower-birds.

Weiske[280] relates a Papuan legend concerning the Fawn-breasted Bower-bird that should be investigated. He says that the natives claim that the 'bower of this bird will not burn'. . . 'If the inhabitants set fire to the steppe [*kunai*] when hunting for kangaroos [= wallabies] and the flame reaches the bower . . . the bird is said to go to a nearby stream, wet its feathers, return to the bower still wet, and moisten it by shaking its damp feathers and so putting out the fire.'

This story is, of course, wildly improbable, but, if true, would seem to show an intelligence such as has never been recorded in any bird.

8

YELLOW-BREASTED BOWER-BIRD

Chlamydera lauterbachi, Reichenov

THE Yellow-breasted ('Lauterbach's') Bower-bird lives in the thick golden *kunai* and *pit-pit* grasslands and at their timbered fringes in upland New Guinea. It was discovered on a tributary of the Ramu River by the German botanist Lauterbach in the nineties of the last century.[230] The plumage of the male is harmoniously and colourfully adapted to its sunny, grassy environment. The head is olive-bronze in hue. The throat is striped and the breast is pale yellow. The back and wings are darker than the head. Many wing feathers are prominently tipped with white. The eyes are brown.

The female is greyish-brown dorsally, and pale buff, with a tinge of yellow below. Its eyes are brown and its bill is horn black. Its feet are described as 'pale olive-slate'.[210] The bird is about $11\frac{1}{2}$ inches long.

The precise range of the Yellow-breasted Bower-bird is unknown and will be so for many years to come. In addition to the type locality, the species has been collected in south-west New Guinea, the Sepik Valley, and not far from the southern coast of Dutch New Guinea (Kamura River). Chaffer[40] has recently made observations in the Wahgi Valley *kunai*-country at 5,600 feet. It has been thought that *C. lauterbachi* is extremely rare, but in the Waghi Valley, towards the centre of the island, it occurs in 'fair numbers', and as great tracts of this tall, coarse, sheltering grass interdigitate the lofty rain-forests of many regions it is probable that soon the known range of the species will be extended in several directions (Fig. 7). The upland *kunai* (Pl. 18) as well as the adjacent rain-forest is subjected to heavy rainfall. Chaffer says that the bird is uncommonly timid. This part of New Guinea is, of course, heavily populated by Melanesians.

The bower and display-ground

Three bowers were shown to Chaffer in January. Two were built in the *kunai* and one in an area of saplings intermingled with the grass. Gilliard[97] found bowers built in reedy *pit-pit* grass. The bower consists of two parallel walls of sticks and fine grass which form a central avenue about 3 inches wide. These walls, unlike those built by other members of the genus, are said to slope outwards so that at their tops a space of about 5 inches occurs. Each wall is from 4 to 6 inches thick at the base and is about 9 inches high. A second striking departure from the typical *Chlamydera* bower is the presence of two additional walls. One of these is built at each end of the bower and so two additional avenues (Fig. 2) are formed. These are some 4 or more inches wide. The end walls are of about the same proportions as those forming the central avenue. A third remarkable feature is lower elevation of the two cross avenues. The pavement of these is about 2 inches lower than that of the central one. The whole closely knit structure rests on a substantial platform of sticks, which is about 3 inches thick and which does not reach to 'any appreciable extent' beyond the bower (Pl. 16).

Display-things consist of numerous bright blue berries about ¾ inch in diameter, bluish stones, and small red berries. Gilliard says that green fruits and 'lumps of hardened clay' are used as well. These are confined within the avenues—no decorations whatever are placed beyond the walls. Some indication of possible zonation of ornaments was observed by Chaffer at one bower. In this bower, too, a few of the blue berries were inserted about half-way up one of the end-walls. Fallen leaves were promptly removed from the bower. The bowers were deserted after the birds realized that they were under observation. No evidence of painting was apparent. No data concerning the bird's display at the bower have been published.

The collector Shaw Mayer shot a male at its bower in October at 5,000 feet. Its testes measured 10×5 and 9×4 mm. Testes of this size in a *Chlamydera* almost certainly contained spermatozoa even though the organs had not reached the peak of spermatogenesis.

The nesting season

Chaffer examined one nest during January. Native information was verified by observation of the sitting bird. The nest was some 6 feet from the ground in a sapling at the edge of a dense area of *kunai* and fern near the bank of a small stream. It was made of coarse sticks built up several inches from a fork in the sapling. It was about 7 inches in diameter and the inner depression was about 4 inches in diameter and 2 inches deep. This was lined with fine twigs. The nest appears to be of more substantial construction than those built by bower-birds inhabiting lower rainfall areas. This is in keeping with a more or less general rule.[182]

Only one egg, partly incubated, was present. This was described by Iredale[124] as being 'a pure oval shape, the ground colour pale sea green, closely concentrically and irregularly streaked with brown of various shades, the streaks thicker, paler and more distant towards the smaller end, the streaks finer, darker and very dense towards the larger end where blotches appear on the extreme end'.

Discussion

The above species is the only member of a typically Australian genus that is at present confined to the great island of New Guinea. It is clearly a *Chlamydera* and more closely related to the Fawn-breasted Bower-bird than to the two species confined to the mainland. Its plumage in the male is obviously an adaptation to the sunny, yellow grass-lands which the species has colonized. The female retains the sombre hues characteristic of the mainland species of *Chlamydera*.

The bower architecture is of singular interest. Allowing for the changed inclination of its walls, the bower is typical of the genus *Chlamydera* except that two end-walls have been added in place of the two display-grounds used by other members of the genus. Fig. 2 shows the typical twin-walled avenue type of bower built by *Ptilonorhynchus*, *Sericulus*, and two species of *Chlamydera*. The *Chlamydera* habitually make display-grounds and accumulate display things at both entrances, as well as a certain number in the central avenue and often a few in the walls. The Yellow-breasted bird, specifically isolated from the Fawn-breasted Bower-bird and the others, has added an extra wall of sticks and an avenue at each end in substitution for the end platforms of sticks and the large display of bones, pebbles, or shells. In the two additional avenues it forms its own characteristic end-display and, in addition, still decorates the original central avenue and sometimes inserts ornaments among the sticks of the walls. Thus it will be seen that this bower, at first sight so different from those of its congeners, is in reality a simple extension of the fundamental architecture. Some bowers (Pl. 16) have very substantial middle walls whilst their end walls—the historically recent modification—remain rudimentary. Mayer[186a] believes that the older birds become 'master builders' and says that their terminal avenues come to resemble 'two sweeping parallel roadways'. It is in these, as we have seen, that most of the decorations are placed, for the 'roadways' take the place of the display-grounds of other members of the genus.

As is the case with *C. cerviniventris*, the display-things of the present species do not resemble the colours of either sex. Chaffer[40] has suggested that the partial choice of blue decorations by the brown-eyed *C. lauterbachi* 'somewhat upsets' the theory that the colour-preference of *P. violaceus* is governed by the physical coloration of the mate. This is not so. The fact that some species of a family exhibit a certain remarkable behaviour form is no valid argument that all should do so.

9

REGENT BOWER-BIRD

Sericulus chrysocephalus (Lewin)

THE male Regent-bird, with its 'brilliancy of the yellow, the richness of the orange, and the unsurpassed depth of the velvety black',[217] is the most beautiful of the Australian bower-birds. It was the first to be described. This is not surprising, for although the Satin-bird has always been the more common in the area of original settlement, exploration parties made the short journey north to the Hawkesbury River rain-forest in the earliest days, and the arresting plumage of the male made its early discovery a certainty. So it was that as early as 1808 it was painted and described by John Lewin, 'a painter and drawer in Natural History'[8, 148] who emigrated to 'New Holland' in 1800. Nobody knows precisely where the original skin came from. Gould heard rumours that the species existed at Sydney. The Sydney area is ecologically unsuited to the Regent-bird, but it is possible that a few still existed in the National Park—Lilyvale rain-forest valleys when the First Fleet arrived. The Green Pigeon, another northern rain-forest species that seems today restricted to country north of the Hawkesbury, has been seen once in the southern valleys mentioned above.[179]

Today the Regent-bird has a continuous, or almost continuous, range from central-eastern New South Wales to south-eastern Queensland in the Blackall and Bunya ranges above Brisbane (Fig. 8). Unlike the Satin-bird, the Regent Bower-bird has no northern race in the tropical rain-forests beyond Townsville.

The Regent Bower-bird was described first as a 'honey-sucker' (Meliphagidae), and later as an oriole. Later still, it was described as having affinities with the birds-of-paradise. The years passed; other bower-birds were discovered and their bower-building habits described. Then, in the sixties, Waller[279 a] in Brisbane and Ramsay[223 a] on the Richmond River found its bower and placed its relationship beyond doubt. Although Lewin described the bird as the 'Golden-crowned Honey-sucker', it would seem that the colonists very soon came to call it by its present name in honour of that fine character, the 'First Gentleman of Europe', whose Regency began in 1811. The Prince Regent became George IV in 1820. In Lewin's next work,[149] published in 1822, the bird was called the 'King Honey-sucker' and not apparently after the local governor of that name as has been sometimes claimed. In 1826 the name Regent-bird appeared in a list of Australian animals[81] and the bird has been so called ever since. The Scottowe MS* dated 1813 further reveals that the bird was named after the regal and not the vice-regal personage. Scottowe, in fact, claims that it was he who first named the bird 'The Regent'. He procured the specimen, he says, on the same day

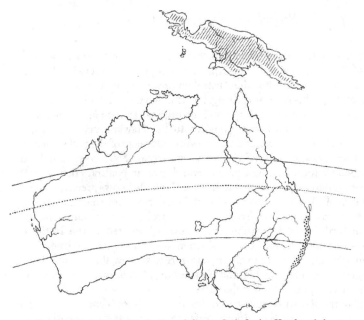

FIG. 8. Approximate distribution of Genus *Sericulus* (= *Xanthomelus*).
Spots: *S. chrysocephalus*.
Stripes: Possible distribution of '*Xanthomelus*' (subject to modification).

that he 'receiv'd in this distant part of the World the News of the Regency Restrictions on His Royal Highness the Prince Regent, having been taken off', and as a small tribute from the Esteem that he bore 'that exalted Character', he called the *Sericulus* the Regent-bird.

The plumage of the male Regent-bird is a uniform velvet black with the exception of its golden head and nape, yellow upper-back and yellow-splashed wings. The plumage on its head is arranged like short, tightly

* 'Select Specimens from nature of the birds, animals, &c. &c. of New South Wales collected and arranged by Thomas Scottowe Esq. The drawings by F. R. Browne, Newcastle, New South Wales, 1813.' The paper in the volume is watermarked 1805, 1807, and 1810 and the Regent plate is watermarked 1810.[157]

packed moss; the feathers begin to lengthen at the nape. The eyes and
beak are yellow and the legs black. The female (Pl. 17) is plain by comparison. It has a patch on the crown that is 'purplish-black'.[185] There is
a smaller dark patch below the throat and another high on the back. The
chest is of a sort of 'honey-comb' mottling and the remainder of the back
is mottled darker still. The wings and tail are brown, and the beak, legs,
and feet are blackish-brown. The skin at the gape is golden-yellow.
The eyes are golden-yellow mottled with brown. The young of both
sexes are drab. Phillipps,[217] who bred the species in captivity, came to
the conclusion that the young male assumes the full adult plumage
during two annual moults when it is 'about four, or possibly not until
rising five'. This writer gives detailed information concerning the
plumage of aviary-bred young, and has also an account[218] of the change
of a bird (which he believed to be a female) into a male plumage in a
single moult.

The Regent Bower-bird is about 12 inches long. Mayr and Jennings[191]
have shown that the adult males have wings, as well as the tail, shorter
than those of immature birds and make the interesting suggestion that
this highly remarkable alteration in wing-shape may be correlated with
courtship performance.

The Regent-bird has a reputation for shyness, but if the observer
remains still it will not infrequently approach within a few feet in the
dense rain-forest. In fact, it has been known to breed in an orange
orchard and to hop into a kitchen where it was fed on scraps.[145] From
earliest times the Regent-bird has raided cultivated fruit-crops,[104] but its
principal food is native fruits and it is particularly partial to the ink-
berries and wild raspberries that grow at the sunny fringes of the rain-
forest. It is known to eat beetles in the bush, and it eats insects vora-
ciously when taken into captivity. In many ways it appears to be a natural
competitor of the larger and numerically more successful Satin-bird in
whose company it sometimes feeds. Although the Regent remains fairly
plentiful in many localities and, probably, was never a common bird, its
numbers have decreased drastically with the march of white settlement
and the felling of the scrubs. It was once a common sight in Australian
bird-dealers' shops, and in 1903 the British market was 'flooded with
them'.[217] The latter statement is doubtless an exaggeration, but certainly
the species was a not uncommon aviary-bird in Edwardian England, and
it was occasionally on the market in Berlin as well.[203]

The display-ground, bower, and associated activities

No laboratory material has been available. Although Regent-birds are
found in the breeding area throughout the year, various accounts suggest
that greater concentrations appear early in July, which is, of course,

about a fortnight after the winter solstice. Bowers (Pl. 17) have been found late in July[179] and in August.[221] Most records of bower-building relate to much later in the season but this may be due to the habits of local naturalists rather than to those of Regent-birds. There is a widespread supposition that the Regent rarely builds a bower, but this belief is probably without foundation. Nowadays the species is restricted to areas that are comparatively remote from cities, and eggs, not bowers, have been the quarry of most observers. Furthermore, the Regent almost always builds its bower not at the outskirts of, but fairly deep within, the rain-forest. In addition, it protects itself by building in a tangle of spiky, gripping, lawyer-vine or 'wait-awhile'. Again, the bower is small. The display is not noisy. Also, the bower is meagrely decorated and the ornaments harmonize, rather than contrast violently with, the jungle floor. The Regent-bird is at first approach shyer than is the Satin-Bower-bird, so if movement is not extremely cautious it is easy to pass a bower by at a range of a few feet without detection. Thus, Ramsay[223 a] was surprised to see a Regent fly to the ground and sit within a yard of where he stood in a Richmond River rain-forest. He remained uncomfortably motionless on a log for 5 minutes while the black and golden bird moved around him. It next hopped to its bower 'scarcely a yard from where I was standing: had I stepped down off the log I must have crushed it'.

The bower walls are made of twigs. They are from 10 to 12 inches high, 2 or 3 inches wide and 7 or 8 inches long. The avenue is $3\frac{1}{2}$ to 4 inches wide. A few of the twigs arch overhead. The whole is built into a platform of twigs which may be about 22×19 inches[36] in extent, but often much smaller. In some bowers the platform does not extend beyond the wall and it is said that some bowers have no platform at all. Most observers comment on the 'carefully swept appearance' of the earth near the bower where no fallen leaves are allowed to remain. In general, the bower itself is not as neat, nor as elaborately constructed, as that of the Satin-bird.

Display-things are generally few. A nineteenth-century author[223 a] described land-shells of five or six species, and blue, red, and black berries which gave the bower 'a very pretty appearance'. Also present were several newly picked leaves and young shoots of a pinkish tint. The date was October. No other bower with such a lavish display has been described. Most have merely a few brown snail-shells, a small number of brown palm-seeds, and perhaps an odd fresh, purplish-tinted, leaf or two. One bower, however, contained a single broken land-shell and twelve glossy black berries, each about twice the size of a domestic pea.[98] Jackson[130] found a bower in a tangle of vines on the Macpherson Ranges on which were snail-shells, fresh five-petalled yellow flowers of the genus *Hibertia*, and fresh yellow leaves, as well as berries and seeds. Pure yellow decorations have apparently not been recorded elsewhere. In captivity, deprived of their natural display-things, Regents have gathered 'smooth round stones about the size of a small marble' and fresh green leaves, and when the English weather inhibited bower-building they placed the above objects in their circular food dishes.[217] Unlike the Satin-bird, but similar to members of the *Chlamydera*, the Regent has usually a few ornaments in the avenue of its bower.

There have been suggestions that the Regent Bower-bird sometimes constructs a simple platform of sticks similar to that built by young Satin-birds, but this has not been proved.

There is no good description of the display of the Regent-bird. Campbell,[36] in November, saw three birds at a bower, two of which he considered were immature males.

They were not in the least disturbed by our presence. One would go into the centre of the bower and, picking up a shell, of which there were three, would dance, half opening its wings and then, tossing the shell in the air or over its head, would run out. While this was going on, the other two birds outside were scraping or sweeping the ground with their wings, and when the shell fell, one would pick it up and enter the bower to go through the same performance as the first bird, and so on. There were four or five fresh leaves in the bower at the time, and on visiting the locality the following afternoon, these were seen to be thrown out and fresh ones placed in their stead.

At a bower in January, Campbell again saw three drab birds at display. 'Each carried an empty shell and in turn went into the bower, and after bobbing up and down a few times with half-opened wings, would toss the shell over the wall. The two birds remaining outside performed various antics and brushed the ground with their wings, as a result of which the soil within the enclosure was quite bare.' During subsequent visits the bower was unoccupied.

Waller[279a] saw a male bird 'playing on the ground, jumping up and down, puffing out its feathers, and rolling about in a very odd manner'. He shot the bird and then found the bower. While he was examining it the female came down; she called loudly in the vicinity of the bower for several successive days. (The female Satin Bower-bird remains near the bower in similar circumstances.) The above, incidentally, was probably the first bower ever found. Goddard[98] heard a 'peculiar low chattering' in dense rain-forest and found a fully adult male Regent at its bower late in October. The walls were unequal in height, one being 6 inches tall and the other a mere 3 inches. There was no platform. The male rearranged the twigs. After it left, the walls were found to be painted with 'a dry yellowish substance, which adhered to the sticks and twigs'. At

9 o'clock next morning both male and female were at the bower. The female flew away, but the male remained.

He wiped the extremity of his bill all over the sticks with a pecking action. I could clearly distinguish a small piece of greenish material held between the mandibles towards the extremity of the bill. Later he picked up a fresh supply of the material from the floor of the avenue and continued operations as previously. Next he turned his attentions to the less robust wall and applied a little of the material. A little later he flew off. . . . Close scrutiny of the walls disclosed that the upper ends of the sticks and twigs were coated with a wet mixture of saliva and a macerated, pea-green vegetable matter.

The bower was deserted shortly afterwards.

There has been speculation as to whether the Regent, like its relatives, is an accomplished mimic of other species. Phillipps says that in captivity it has a 'pleasing little song'. 'Squatting in some sheltered corner, softly and sweetly he will warble away for quite a long time; and he is no mean mimic.' Phillipps says that he heard a female singing like a male. Recently there has come confirmation of the former observation. O'Reilly[212] saw a male Regent sway its body and mimic, 6 feet from the observer, the territorial cry of the Satin-bird and two other species in 'midsummer'. The mimicry was included in a whisper-song that would probably not be heard at a distance of several yards. There was no bower, as far as was ascertained, in the immediate vicinity.

Aviary observations

A series of highly diverting observations were made on captive Regent-birds in England early in the present century by Phillipps,[217] who said that the females build what he first called 'love-parlours' and later 'nuptial bowers'.[219] Such a bower has never been found in the bush but, in view of the secluded habitat of the Regent, this does not mean, perhaps, that it does not exist. It will be ironic if there exists still unknown in Australia a type of bower that was described in London nearly half a century ago. Phillipps indiscriminately mixed his Regent-birds with quail, Long-tailed Whydahs, and other species, but they built bowers, displayed energetically and reared one brood in a suspended basket filled with straw! His accounts contain some of the most valuable data yet recorded. Unfortunately it is mixed with a lot of anthropomorphic nonsense of a kind that pleases bird-fanciers but which renders suspect some of the phenomena he purports to describe.

The female bowers were described as follows:

*These love-parlours, each one built by a female for her sole use,** rather open and not far apart, and each most jealously guarded by its fair owner, were of the

---

* Original author's italics.

shape of a horseshoe magnet but with the sides equidistant throughout their entire length, open only at one end, and inside of about the same length and breadth as the bird, the top of the barricade being about on a level with the back of the squatting female, the sticks woven together, being laid flat, none upright. The female would enter and squat in her love-parlour, the tail remaining towards the entrance, whilst a male, with every imaginable and unimaginable contortion, accompanied by a continuous discharge of (vocal) fire-arms, would make rushes and furious (feigned) assaults on the front of the breastwork, the female sitting in a lump and not moving a muscle. Every now and then, however, the male would slyly work round to the rear and tweak the tip of the female's tail. This advance, at any time perhaps but the early morning, or at any time when I was looking on, was not considered correct, and the female would slowly turn her head with what we will suppose was an icy look of grave disapproval.

Another female (in the autumn) was

the most energetic, and her fortress became a really formidable structure, the parapet being raised *pari passu* with the additions to the platform. Every bird that approached was savagely driven away with the exception of a female Rain Quail who sat on her eggs and brought off her brood scarcely a foot away and was never molested—not until the chicks had come out were the shells pounced upon. . . .

The foregoing seemed to be the recognized courting arrangement, the selected female, with or without her protecting barricade, squatting lumpily on the ground, on the proposed nesting-site, and in the nest itself, while . . . the male sported before her. On the other hand, the rejected female, at first favoured but afterwards deserted owing to an accidental circumstance which will be narrated later, built or partly built three love-parlours in different spots in her endeavours to bring back to his allegiance the fickle mate.

If there is a superabundance of females, or if the species be polygamous, these endeavours on the part of the females to entice the males may be readily understood.

Phillipps says that

the ordinary bowers [= twin-walled structures described in the bush] are used more frequently by the males, not as 'runs', but for the purpose of showing off before one or several females perched close by and above them. Viewed from below, only dense black is exposed to the sight, so they perch high while desiring to conceal themselves. Viewed from above, rich orange and brilliant yellow meet the eye, and so, when the male wants to display before his females, down to the ground he goes and there disports himself. If displaying elsewhere the rule holds good, for he bows his head to the level of the perch, so as to show off the orange and yellow of head and back. As they go through their extraordinary dances and performances, they constantly look back over their shoulders, backwards and upwards, as if appealing to their lady loves with a 'Wasn't that fine?' kind of air; and all the time be it remembered they are clattering and chattering in an idiotic manner in low and subdued tones.

No pen [says Phillipps][216] can describe the antics of the bird when courting in the vicinity of his bower; they are wonderful beyond description. . . . All the Regent's antics are accompanied by indescribable notes and noises, not loud but all expressive and quite in keeping with the general weirdness of the performance.

Phillipps declares that a second female displayed to the male after the first female went to its nest. The male also displayed to the female. She attempted to build a nest in the autumn (October), but ovulation did not occur. This apparent pointer towards polygamy, however, may have been induced by the unsatisfactory conditions in which the birds were kept. It is perhaps significant that Phillipps's male 'paid homage' to a female Thrush 'to the intense disgust of his long-suffering mate'.

The nesting season

The first nests are generally found early in November. It would seem that the female alone constructs them. Gilbert,[93] incidentally, saw a female bring nest-material every 3 minutes during one period and observed that the male kept strictly away. Fresh eggs have been recorded in January,[36] but this appears to be uncommonly late. The nest is concealed in the dense jungle, generally amid a tangle of vines at a height of 12 to 40 feet. It is a saucer-shaped structure like those of all other bowerbirds. Loosely built of twigs, it is about 10 inches in diameter with an egg-cavity some 4 inches wide and 2 inches deep. This chamber is sparsely lined with twigs or plant-stems. Sometimes the eggs can be seen from below.

Two or occasionally three eggs are laid. They are elongate or swollen oval and have a ground colour that ranges from off-white to greyish- or yellowish-stone. They are streaked by hair-like markings of umber-brown or purplish-black which swerve and loop in all directions and encircle the egg. Sometimes the eggs of *Sericulus* are barely distinguishable from those of *C. maculata*.

The female in captivity is said to have 'a suppressed ventriloquial note' when near her nest. The incubation period in captivity is perhaps 18 or 19 days. Phillipps has further information of high interest, especially that concerning the feeding of the young. We have seen how vital is a large supply of easily captured protein food for the survival of the young of the Satin-bird. In captivity, the female Regent refused to take fruit, her own principal diet in the bush, to the young. Phillipps had anxious times, as had Hirst with the Satin-bird, procuring sufficient animal food to keep the young Regent-birds alive. The female Regent did not like meal-worms, but accepted cockroaches whenever they were available. These she always took to the nest, although adult Regent-birds are themselves especially partial to this particular insect. She only briefly took egg-flake and sultanas to the young, but was, surprisingly, more prone to use 'biscuit-sop'. 'During this time she did what I never before saw a Regent do—dart into the air after any winged creature that might chance to enter the aviary.'

Chaffer[41] has published brief but interesting observations on feeding habits in the bush. A young bird about to leave the nest was fed on wild raspberries. But a week previously moths, caught on the wing, had been used. It will be recalled that Hirst's Satin-bird chicks died in the first two weeks of life when the captive female was unable to provide them with sufficient protein food.

In wild birds an unmistakable post-nuptial flocking phase occurs. The usual stories of 'one coloured male to about 50 hens' are told; but as many as 3 or 4 golden birds have been recorded in a flock totalling only a dozen birds altogether.[35] This proportion, however, is unusually high. Most flocks of 10 or 20 seen by me contained only one or two brilliantly plumaged males. Occasionally a young male in the process of colour-change may be seen. The flocks travel for unknown distances in search of fruit but, unlike Satin-birds, Regents never venture far away from the coastal rain-forests. Some flocks appear never to move more than a few miles from the breeding area. In any case, they are generally back there in July.

Discussion

The Regent-bird was described first as a honey-eater, next as an oriole, and later as a bird-of-paradise, and it was not until it was seen at a bower that it was unquestionably accepted as a bower-bird. Even to-day, on the mere consideration of surface anatomy, the species is not generally regarded as being closely allied to either *Ptilonorhynchus* or *Chlamydera*. Mathews,[185] for example, has said that it is of 'apparently no relationship with the other birds with which it is associated mainly on account of its bower-building habits'. Yet it builds a bower of the same avenue pattern as the others, brings decorations to its display-ground, and uses them there in its posturings before the female. Furthermore, it shares with the Satin and Spotted Bower-birds the fantastic habit of painting its bower. If this habit is an extension of the widespread avian ritual of courtship-feeding it is possible, though unlikely, that its development in the three above genera is the result of convergent evolution. At present, however, I believe that, along with the construction of the avenue-type bower and the display to the female nearby or within it, bower-painting is an inheritance from a common ancestry. The seasonal movements, the nest, and the eggs of the Regent-bird are all extremely similar to those of various species of the two other genera mentioned above. The striking way in which a morphological character can change is evidenced by the surprising discovery of Mayr and Jennings that *Sericulus chrysocephalus* has smaller wings and tail in the adult male than in the females and juveniles. It will be of great interest if the above authors are correct in their surmise that these characteristics are an adaptation to some kind of bower-display. It will be recalled that the bower surroundings are usually swept, and one observer says the wings are used for this purpose.

The female Regent-bird is essentially black and brown but it has yellow, brown-mottled eyes and the skin at the gape is yellow. It will be recalled that of the almost numberless kinds of potential display-objects within its environment the bird chooses always brownish snail-shells, and often brown palm-seeds. Black berries have been recorded and, on one occasion, yellow objects.* The only total departures from the colour scheme of the female are blue and red berries mentioned as bower-decorations in a single account of 90 years ago, and the green and tinted leaves that do indeed seem to occur on many bowers. It will be recalled that although the Spotted and Great Grey Bower-birds select sun-bleached or otherwise pale or highly reflective decorations that more or less match the plumage of the females, they, too, still utilize green leaves, berries, and occasional plant-galls.

In *Sericulus* the correlation between the colours of the display-things chosen by the male, and the plumage of the female to whom the display is directed, is nothing like as striking as in the Satin-bird. But if we consider the almost unlimited range of articles at the males' disposal, we cannot, in the absence of experimentation, yet deny that such a correlation exists. Such experiments will be carried out some day. Further, a cinematographic record of the bower-display of members of the three avenue-building genera should, in some future analysis, throw considerable further light on their relationship.

It will have been noticed that the bowers of *S. chrysocephalus* are much smaller and the number of decorations fewer than any other species yet dealt with. This has led to the opinion that the Regent-bird is 'still a novice at bower-building',[124] that it may be 'only learning the art', and that in it bower-building is a 'possibly recent habit'.[126] There is no evidence whatever to support such suppositions. The Regent-bird's bower, however small, is apparently an adequate vehicle for its sexual display, and such being the case, there is no reason for it to be any bigger or more lavishly decorated than it is.

* See Addenda, note 'B', p. 190. It is possible that these colours are chosen because they match the old black and gold males as well as the brownish young ones.

10

NEW GUINEA REGENT BOWER-BIRD
Sericulus (Xanthomelus) bakeri (Chapin)

THE discovery of a new species of bird in the biologically unexplored hinterland of New Guinea is never surprising. The present bird, however, was collected by Beck[44] as recently as 1928 near Madang on the northern coast in a region that has been patrolled for decades. It is astonishing that such a brilliant animal was not described before. It must have been handled by Germans and Australians during the plume-trading era before, and immediately after, the First World War, for the mountainous country behind the Rai Coast was a most prolific area of slaughter. Such a distinctively coloured bird would certainly attract native hunters. However, such was the craze for skins with elongated plumes (birds-of-paradise, egrets, and, to the lesser extent, *guria*-pigeons), that the present beautiful bower-bird was apparently not thought worthy of close inspection, let alone shipment to the plume-houses of Hollandia, Batavia, and Europe.

The male broadly resembles the Australian Regent on casual inspection. The short crown-feathers, however, are scarlet and, unlike the Australian Regent, there is a fiery orange-scarlet dorsal 'cape' or 'ruff' composed of feathers that reach a length of over 2 inches, and which may be partially erectile. It has black and yellow wings very like those of the Australian bird, but its black ventral surface has possibly not the velvety 'midnight' intensity of that of *S. chrysocephalus*. The female of *S. bakeri* has not been described. The immature male is brownish above and its breast and flanks are cream-buff, irregularly barred with blackish-brown. The species is about 12 inches long. The precise altitude at which the three Madang skins were taken has not been recorded and the range of the bird is unknown.

The habits of the New Guinea Regent-bird are totally unrecorded. This is regrettable, but not surprising when we reflect how little is known of the behaviour of the Australian Regent, a bird that occurs within 60 miles of a city of $1\frac{1}{2}$ million people.

11

GOLDEN-BIRD

Sericulus (Xanthomelus) aureus (Linnaeus)

THE male of *S. aureus* is startlingly beautiful. It has a golden-red, sub-crested head and an extensive and a possibly erectile dorsal mantle or 'cape' of elongated feathers of the same colour (Fig. 9). The feathers of head and mantle glint in the light. It has yellow, black-tipped wings, a yellow back, and a black, yellowish-tipped tail. Its throat is black. The rest of the ventral surface is bright yellow. Its eyes are yellow. Its beak has a yellow base, merging into black. The female, too, is prettier than those of other bower-birds. She possesses neither crest nor mantle, but she has a yellow ventral surface, a paler throat, and an olive-brown head, wings, and back. The immatures are broadly similar to the female. The bird is about $10\frac{1}{2}$ inches long.

The plumes of the Golden-bird, along with those of the various birds-of-paradise, have been used for untold generations as Melanesian head-gear and regalia of the *sing-sing* or ceremonial. As a native ornament and item of trade, it was bartered to early European navigators long before the Australian mainland was settled. Skins reached Europe with those of birds-of-paradise, and as a consequence the Golden-bird was the first bower-bird described. It was called the Golden Bird-of-Paradise by Edwards in his *Natural History of Birds* published as early as 1750, and Linnaeus described it as the Golden Oriole, *Oriolus aureus*, in the 10th (1758) edition of his *Systema Natura*.[150] The type locality was given as 'Asia'. It was Linnaeus, incidentally, who named the Greater Bird-of-Paradise *Paradisea apoda*, using the specific name in allusion to the old travellers' tales that these birds, living and breeding in paradise, did not require feet. Most of the skins which percolated through to Europe in the sixteenth, seventeenth, and even eighteenth centuries did, in fact, lack feet. This was because native taxidermy is of the crudest possible kind. A ventral incision is made and the skeletal structures and viscera are removed. A piece of bamboo is swathed in fibre and thrust into the body. The 'mounted' skin is then sometimes smoked over a fire to help preservation. The legs are usually cut off altogether. The aim of the natives is to show off the most beautiful feathers. If the wings are not attractive they are sometimes cut off too.

By the time the French surgeon Lesson[146] brought back his famous collection, bartered from the natives of north-west New Guinea, the Australian Regent Bower-bird had been described and so Lesson placed the Black-throated Golden-bird in the genus *Sericulus*. Later the name *Xanthomelus* was proposed for it.[27] In 1858, 40 years after Lesson, Alfred Russell Wallace was engaged in his classical researches in the East Indies and made a special journey from Ternate to north-west New Guinea in search of birds-of-paradise. The Golden-bird was among the few brilliant birds that he managed to get. Wallace did not believe that

FIG. 9. Male Golden-bird (*Sericulus aureus*).
(*After Sharpe.*)

it was a *Sericulus*. 'The form of the bill and the character of the plumage seem to me to be so different that it will have to form a new genus', he said.[279] From now on most people referred to the species as *Xanthomelus aureus*. The great Italian traveller D'Alberti secured it at an altitude of 3,000 ft. at Arfak. Natives told him that it 'nested' on the ground, but neither D'Alberti nor his fellow countryman Beccari found the terrestrial 'nest' which was, no doubt, the bower. Of more recent years several collectors have brought the bird to Europe and the United States, but its exact range is still unknown. Specimens have been taken in the Vogelkop, Onin Peninsula, and the mountains behind Geelvinck Bay.[188] It lives in the rain-forest to an altitude of 4,200 feet, and feeds largely on figs. Its bower, with certainty, has never been found.

The Golden-bird is represented in south-eastern New Guinea by an equally beautiful yellow-throated sub-species which was described as *Xanthomelus ardens*. D'Alberti bartered the first known skin from a

Papuan on the Fly River in about 1875.[61] The male of this race has an even more fiery orange-red dorsal surface, including the elongated mantle, than its black-throated relative. It has also a dark crimson streak around the eye. Significantly, a small portion of the plumage around the eyes is of the same short, close-packed, moss-like character that is such a striking feature of the head of the Australian *S. chrysocephalus*. (This is a somewhat unusual *vestigial structure*!) The ventral surface is yellow, tinged with orange. The wings are orange-yellow, margined with brown, and the tail is blackish brown. The female is not unlike that of *S. a. aureus* except that her ventral surface may be paler.

The Black-throated Golden-bird has since been found west of the Fly River, far into Dutch New Guinea, and at Mount Hagen in the central highlands of Australian New Guinea. Nothing is known of its habits.

Discussion

We have dealt with a group of New Guinea birds whose affinity with the Australian Regent-bird has been always disputed. It is unfortunate that the bower of none of them has been described. Shaw Mayer[186] was shown a bower by natives who claimed that it was built by *S. aureus*. 'It resembled the bower of the Australian Regent-bird but a little larger. The walls were straight and there were no decorations. It was built under a stand of bamboo . . . at about 4,000 feet.'

The description of '*Xanthomelus*' *bakeri* seems to remove any doubt that this group is in fact closely related to the Australian Regent-bird for, as has already been stated,[44, 191] the colour-pattern of the bird is that of the Regent whilst the form of its beak and its elongated mantle feathers show its affinity with the Golden-birds of New Guinea. Before the discovery of *S. bakeri* it was not surprising that many people preferred to keep the Golden-birds in a separate genus, *Xanthomelus*. Iredale, however,[124] still asserts that, despite the many similarities between the Australian Regent-bird and the New Guinea Golden-birds, it is doubtful whether the two groups are closely related.

It is of interest that more than half a century ago both Salvadori and Sharpe[256] suggested that when the bowers of the Golden Bower-birds were discovered, it would probably be found that they build 'bowers' (= avenues) and not 'huts' (= maypoles). If the New Guinea Golden-birds are in fact generically similar to the Regent-bird of Australia, it is highly probable that they will be found to build avenue-type bowers of the sort indicated by Mayer's native informants. In view of the differences among systematists as to the relationships of the *Sericulus*/*Xanthomelus* group, the discovery of the bowers and display habits of the New Guinea birds is of highest possible priority.

12

BROWN GARDENER

Amblyornis inornatus (Schlegel)

THE publicity given by Gould to the Australian bower-birds and their habits at about the middle of the nineteenth century led to highly romanticized accounts in popular magazines, and fairly soon the expression 'bower-bird' came to be applied to people who accumulated useless odds and ends. In 1878 Victorian English households received, in *The Gardener's Chronicle*,[17] the further astonishing news that there existed in the virgin forests of Dutch New Guinea a bird that made not only a collection of objects that are 'very various, but always of vivid colour', but, in addition, built a hut and, in front of it, planted an 'elegant little garden' (Pl. 19). So the Gardener Bower-bird came to be described in English. This account was a translation from the original Italian of Beccari, and is still the most extensive treatment yet published.

Reports of the strange 'nest' of a rather undistinguished-looking bird of the Arfak highlands had been brought to the coast by plume-hunters employed by Dutch traders and planters. The bird itself was worthless as an item of trade, for its feathers were merely 'of several shades of brown' and showed 'no sexual differences'. It was classified as a bower-bird, and named *Ptilonorhynchus inornatus* by the Dutchman Schlegel in 1871.[249] Beccari heard of the bird and its 'nest' before he went inland. At 4,800 feet, on a spur of Mount Arfak where the forest was so dense that 'scarcely a ray of sunshine penetrated', he came suddenly upon a conical 'hut or bower close to a small meadow enamelled with flowers...'. He gave strict orders to his collectors not to destroy 'the little building' but this was an 'unnecessary caution, since the Papuans take care not to disturb these nests or bowers, even if they are in their way'. The natives call the bird *Buruk Guria*, the 'Master Bird', because it 'imitates the songs and screamings of numerous birds', and *Tukan Kobon*, 'which means a gardener'.

Beccari took his colours, brushes, pencils, and gun, and went back to sketch the bower (Pl. 20). It was built on a flat space with a sapling as thick as a walking-stick for a central pillar. At the base of the sapling the bird had constructed a kind of cone, chiefly of moss. Beccari's illustration shows that this cone does not support the hut, or cabin (*capanna*).

The roof of this slopes up from the ground to meet the supporting column at a height of about 2 feet. A front entrance leads to an interior passage which encircles the cone. The hut is waterproof and—a remarkable fact—is built of the thin stems of *Dendrobium*, an epiphytic orchid which grows prolifically on the surrounding tree-trunks. In the perennially damp rain-forest, this plant continues to live and therefore does not moulder.

'Before the cottage', wrote Beccari, 'is a meadow of moss' which is transplanted there by the hut-builder.

On this green, flowers and fruits of pretty colour are placed so as to form an elegant little garden ... it would appear that the husband offers there his daily gifts to the wife. The objects are very numerous, but always of vivid colour. There were some fruits of a Garcinia like a small-sized apple. Others were the fruits of Gardenias of a deep yellow colour in the interior. I also saw small rosy fruits ... and beautiful rosy flowers. There were also fungi and mottled insects placed on the turf. As soon as the objects are faded they are moved to the back of the hut [= outside the actual structure].

'The passion for flowers and gardens', said Beccari, 'is a sign of good taste and refinement'. He regretted that the human inhabitants of Arfak did not emulate the bird. 'Their houses are quite inaccessible from dirt', he said tersely. Beccari called it the 'bird gardener' (*gjardiniere*) adding that its specific name *inornatus* (given it by Schlegel) well suggests its very simple dress. The sexes, as Beccari said, are externally identical. The bird has a reddish-tinged brown dorsal surface. Below, it is yellowish or orange-tinged brown. Its beak is black, its eyes chestnut-coloured, and its legs are grey. It is a little over 9 inches long.

According to Mayr,[187] who collected in New Guinea, the bower-hut is built by a single male. The American collector, Ripley,[234] brought back valuable information just before the last war. 'In one place on the gentle slope, the undergrowth had been neatly cleared away from an area 4 feet square. The place looked as if it had been swept with a broom.' The hut was about 3 feet tall and 5 feet broad at the base (Pl. 19). It was built around a central sapling, but Ripley does not mention the basal cone. The opening was about a foot high.

This curious structure fronted on the cleared area. The impression of a front lawn was heightened by several small beds of flowers or fruit. Just under the door there was a neat bed of yellow fruit. Further out on the front lawn there was a bed of blue fruit. At the bottom of the lawn there was a large squarish bed of pieces of charcoal and small black stones. A few brownish fruits lay here also, some of them rather decayed. Off to one side there were several big mushrooms in a heap, and near them were ten freshly picked flowers.

Ripley had, he says, known vaguely that there were such things as bower-birds in New Guinea, but he was startled by the scene he described. 'The effect of all this', he wrote, 'was overwhelming.'

Later he heard unfamiliar chuckling calls and saw a Brown Gardener in a tree near its bower. 'With a series of rather angular hops the bird sprang from branch to branch until it was on the ground.' It looked around, noticing a match that had been carelessly thrown into the middle of its clearing. It hopped over, picked the match up and with a toss of its head threw it out of the area. 'Then, with a flirt of its tail, it leaped into the house and disappeared, making soft growling churring noises all the while. After a minute it came out and flew off again.'

Ripley collected the pinkish blossoms of a jungle begonia, some small yellow flowers from a vine, and a single pretty red orchid. All these he placed on the middle of the lawn. A single bird reappeared and flew directly to the pile of alien blossoms. The yellow flowers were removed first. After some hesitation, and a good many nods and looks and flirts of the tail, the begonias were removed.

Last of all the bower-bird picked up the pretty red orchid. This time it seemed to be in two minds. It would hop this way and that from one pile of fruit or flowers to another. Finally, with many darts and flourishes the orchid was placed on top of the pink flowers. The two colours swore a bit, but under the circumstances it was certainly the best matching job that could be done.

Ten days after his arrival, Ripley came upon a display-ground that contained several beds of fruit, one big bed of charcoal and, in the centre, six used 0·410 shotgun cartridges. These were red with bright brass butts. Three weeks after his arrival, Ripley paid a last visit to this bower. 'As I came near, I could hear their harsh calls, now long and chattering, now loud, almost like a barking dog. They stopped as I came to the bower and flew quietly away. Where before there had been six shot gun shells, there were now 29 in a sprawling pile which almost covered the entire front lawn.' The above observations were made in March, just before the beginning of the north-west monsoon season, at an altitude of 5,000 feet in the moss-forest at Bon Kourangen, inland from Wejos. The bird's range may be restricted to the mountains of north-western New Guinea.

Discussion

The bower described above is the first of the category of 'maypole' bowers so far encountered. We see in it a radical departure from the avenue-bowers built by members of the genera *Ptilonorhynchus*, *Chlamydera*, and *Sericulus*. Although the bower of *A. inornatus* at first sight most conspicuously suggests a hut with a front garden, I propose to show that the much publicized hut is not the fundamental feature of the structure. Fig. 2 reveals that, from an evolutionary point of view, the primary features of the bower are the central column and the fabric cone packed around its base. It will emerge that this 'maypole' is just as fundamental among the next group of species as are the basic twin walls among the avenue-builders.

13

ORANGE-CRESTED STRIPED GARDENER
Amblyornis subalaris, Sharpe

IF the discovery of the hut and 'elegant little garden' of *A. inornatus* caused something like a sensation, the same certainly cannot be said of that of the far more beautiful Orange-crested Gardener. Furthermore, because the first specimen collected was that of the rather plain female, the bird lay for months in a museum cabinet before it was described. This skin was collected by Andrew Goldie in the Astrolable Range of the Owen Stanley Mountains behind Port Moresby in 1884. It was one of a collection of birds sent to England, a few of which were bought by the British Museum. 'Pressure of work . . . during the removal of the zoological collections to Kensington has not allowed me the time to study them until recently', wrote Bowdler Sharpe in 1884.[255] Sharpe was not sure whether the bird was a new species. 'I at last gave it a name', he wrote 'little dreaming that in the following year the male bird would be discovered, and would turn out to be such a beautiful and striking form of bower-bird.'[256]

The golden-crested male was collected by a Hessian gold-prospector named Hunstein, who had been previously associated with Goldie, and who later ascended the Owen Stanleys alone except for one native 'boy'. Hunstein found himself in 'a world of new trees and plants', including 'a rhododendron with wonderful white flowers' which was later described by von Mueller in Melbourne.[82] Hunstein saw the feather ornaments worn by the Papuans and believed that if he persisted in his search through the dripping uplands he would discover new birds. His male *A. subalaris* was adorned with 'a deep, orange-red erectile hood . . . bordered on the sides with dark brown [with] dark tips on some of its middle feathers'. This hood or crest has its insertion in the crown above the eyes, not directly behind the beak. It reaches a length of some $2\frac{1}{2}$ inches. The back and wings of the male are brown. The undersurface is paler olive-brown, with prominent stripes on the throat and breast. The female is very like the male, except that it lacks the crest and has a less prominently shafted under-surface. The eyes of both sexes are brown, the beak is dark-brown, and the feet are dark olive-green. The bird is a little over 8 inches long.

As far as is known, the Orange-crested Gardener is restricted to the mountains of south-eastern New Guinea, but it may occur elsewhere. It has been found as high as 9,000 feet. Goodwin, who collected with the Macgregor Expedition, found a bird at 6,000 feet on Mount Belford that had an almost completely orange crest which was devoid of the dark margin previously described. It was Goodwin, too, who furnished the first description of its bower.

Goodwin[99] was in camp on Mount Belford when a colleague, Belford, came in with an Orange-crested Bower-bird. Belford urged Goodwin not to leave the locality without seeing the display-ground that he had found.

I am glad to say I followed his advice. At a short distance off, the bower from the back looks like a cart-load of sticks rounded on the top. On going round to the front I saw the most beautiful building ever constructed by a bird. . . . The edifice was dome-like, only half covered over, and exposed to view inside a ring or circus. In the centre of this was built a bank of moss, decorated with flowers and seed, out of which grew a small tree interlaced with sticks. (Fig. 2.)

Simson,[259] who collected *A. subalaris* between 3,000 and 6,000 feet in the Owen Stanleys in the early years of the present century, found six bowers which he attributed to the species and gives a fuller description.

Each consisted of a dome-shaped mass of twigs, about two feet in height and three in width. In this mass there were two rounded openings communicating within and facing the yard in front. Situated between the two openings is an almost black flower-bed, composed of fibre taken from the stems of tree-ferns. Into this bed the bird sticks flowers, berries, bright-coloured leaves and beetles. In front of this structure is a yard enclosed with twigs and over this yard in every playground that I saw were strewn brilliant scarlet fruits, and sometimes a few flowers.

Simson reports that one bower possessed no such 'yard'. In front of this atypical bower there was a forked bough suspended with one limb penetrating the fore-part of the bower. The bough was held in position by the other limb 'which was fixed to the trunk of a neighbouring tree by some sort of extremely tenacious glue'.

Simson seems to have examined this astonishing piece of constructional work with some thoroughness. 'I should think', he says, 'that the united efforts of several birds must have been necessary to fix the bough in position.' He believed that the birds used the bough 'for perching on when going in and out of the tunnel'. Unfortunately the bough was not photographed, although two of Simson's bowers are illustrated[259] (Pl. 20). He goes on to describe the 'flower-bed'. It included pale mauve-coloured beetles, bright yellow trumpet-shaped flowers, small cream-coloured flowers, bright blue berries, and yellowish-green leaves. He commented on the very striking appearance of the decorations, and says that in some display-grounds the decorations were arranged in a definite order. 'In one there were yellow flowers on one side and blue berries on the other.' After the scarlet fruits began to decompose they were removed and left to rot a yard or two away from the bower.

Weiske,[280] who 'often watched' *A. subalaris* building its bower, gives a description similar to those above. He says the birds were found at about 4,000 feet and that bowers are built on mountain slopes from 60 to 150 yards below the ridges. A sapling about an inch thick is selected, and surrounding it is built a basal cone of twigs which is packed with moss. The tunnel, says Weiske, is 'round in cross-section' and is lined with moss.

The structure is adorned on the outside [= of the cone?] with white, star-shaped blossoms, brightly coloured beetles' wings, and hard, shiny blue berries stuck into the moss. The white stars on the green moss make a beautiful picture. In front of the opening of the tunnel a playground is . . . decorated with yellow and other coloured fruits and seed kernels, and with yellow blossoms. The blossoms are renewed as they begin to fade.

Weiske says that the egg is yellowish white in coloration. The nest is cup-shaped.

Discussion

It is almost certain that the bower and display-ground described by the authors above is that of *A. subalaris*. This requires confirmation, however, as does especially the report that the bird sometimes uses a kind of glue with which it fixes a display-branch into position. In the past, most authors have been so obsessed with the importance of describing, to the last spot, streak, and millimetre, the plumage and other parts of the bird that they have paid little or no attention to the equally important things that the birds do. In older writings some bowers described are attributed to species that build quite different structures. In a modern account[126] the question has been further confused by the uncritical inclusion of the original descriptions of two apparently different bowers under the heading of *A. subalaris* (see p. 276).

The bower described above is clearly a maypole-type of the same category as that of *A. inornatus*. Each is supported by a central pole and has a fabric cone surrounding its base. *A. subalaris* apparently builds a stockade around the 'yard' fronting its hut but *A. inornatus* does not.

14

YELLOW-CRESTED GARDENER
Amblyornis macgregoriae, De Vis

THE Yellow-crested Gardener was collected by the Macgregor Expedition in 1889 near what was to become the Kokoda Track, of bloody memory, across the Owen Stanley Ranges. It was shot on Mount Musgrave, and Goodwin,[99] a bushman with the expedition, proposed to call it *A. musgraveianus*. However, a description by De Vis was published before that of the collector. With very proper, not to say humble, respect to the administrator to whom he was reporting on the collection, De Vis wrote: 'A second new Bower-bird, constituting a third species of the genus *Amblyornis*, and distinguished by a very ornate crest, will, if allowed, be honoured with the name of Lady Macgregor. . . .'[68] De Vis, without doubt, had an eye towards future favours, which he duly received; it would have been better, however, had the bird been called after one of the men who sweated over the ranges to collect it.

The male (Fig. 10) has a lustrous olive-brown upper surface. The ventral parts are yellowish-brown. The crest or hood is deep yellow, with an orange-red tint, and is more than 3 inches long. The crest is faintly streaked, and distinctly tipped, with brown. The female has a slightly warmer tint of brown on the dorsal surface than has the male, but, apart from lacking the crest, is otherwise generally similar. The eyes are brown, the bill brownish-black, and the legs are black. The bird is nearly 10 inches long. Blood has recently discovered in the Hagen area a bird which Iredale[126] considers may be a new sub-species. It has a short beak, a very long crest, a yellowish-brown throat, and a dark olive-green breast which contrasts vividly with the pale rufous olive abdomen. Other sub-species have been described from the Hertzog Mountains, and from the ranges on the Huon Peninsula on the north coast of the island.[188] The bird has been found as high as 11,000 feet.

Goodwin said that he had 'never seen a more beautiful bower or playground' than that built by the Yellow-crested Gardener. It was built of moss packed evenly around a small tree to a height of 2 feet and looked 'at first sight like a marble fountain'; small sticks projected from various positions and gave 'the idea of jets of water'. Around the base of the moss-decked tree ran a circus, the floor of which was smooth and even.

This was enclosed by a low parapet (Fig. 2). De Vis published a more informative description:

Around a young tree . . . a circular mass of sticks intermingled with moss, and 45 inches in diameter, is built with perpendicular sides to a height varying from nine inches to two feet. . . . A circular channel nine inches in depth is left

FIG. 10. Male Yellow-crested Gardener (*Amblyornis macgregoriae*).
(*After a photograph by Gilliard.*)

between the tree and the outer edge of the pile [= parapet]. The outer wall of the channel is 9 inches in breadth. . . . The depressed channel is the playground of the bird; in it several individuals of both sexes may be seen pursuing each other round and round. Mr. Kowald, who saw one of these bowers being built, states that all the materials are selected and conveyed by the males to females, who alone are the architects. Mr. Kowald is strongly of the opinion that the increase of the bower in height is almost constantly in progress, and that each is used by the same birds for years in succession.

Chaffer[43] described and photographed a similar bower on the Benimbla Plateau at 7,000 feet in the Hagen area (Pl. 21). The display-ground was in heavy jungle. Only one shy male, with crest folded along its back, was seen.

In this bower sticks were built around a small sapling forming a rough cylinder about three feet in height and eight inches in diameter. A portion, about three or four inches of the base of this structure, was packed with green moss. Hanging from a number of twigs at a height from the ground convenient for the bird to reach, were small bundles of the borings of wood-boring insects fastened together with spider web or similar material. Other than the moss, this [? insect frass] was the only decoration used. Surrounding the central structure was a 'circus ring' of cleared ground, and beyond that an outer ring of moss a few inches in height. The total area occupied by the structure was about three feet in diameter.

Simson,[259] too, described and sketched a very similar structure that he saw in the Kagi area of the Owen Stanleys at about 6,000 feet. It consisted of a space of about $4\frac{1}{2}$ feet in diameter, shaped like a saucer and lined with moss. 'In the centre of the saucer and round the stem of a bush is a loose bundle of twigs with no openings and about a foot and a half in height.' (Pl. 21.) He was unable to say which species built it.

Weiske[280] described a Papuan bower belonging to a bright-crested bird that he called *A. inornata* (the uncrested Brown Gardener from Dutch New Guinea) but which probably belonged to *A. macgregoriae*. Weiske's bird lives at a height of 6,000 feet and is called *Lorli* in the Koiare district and *Golowali* by the Aroa River people. It constructs its bower on a ridge 'by spreading a wide ring of moss of about a metre in diameter around one, and sometimes even two, saplings . . . a kind of frame of withered twigs and branches is then built into the branches and twigs of the trunk to a height of $1-1\frac{1}{2}$ metres. . . I have never seen a bower of this bird that was furnished with any kind of decoration.'

A series of skins of *A. macgregoriae*, taken by Shaw Mayer at about 5,000 feet altitude in north-eastern New Guinea, shed interesting light on the sexual season of the species. Mayer, who sketched on a specimen label a bower resembling that described by Goodwin, Chaffer, Simson, and Weiske, made a careful outline sketch of the size of the gonads on the label of each bird collected. A female taken on 29 May on the Purari-Ramu Divide possessed only minute oocytes. A crested adult male taken on the same range at about the same time had both testes measuring $2 \cdot 2 \times 1$ mm. In September a crested male taken at 6,000 feet in the Upper Waria River showed testis enlargement to the extent of $3 \times 1 \cdot 5$ mm. and a second adult male of the same month had testes which measured 6×2 mm. Spermatogenesis, at least to the degree of primary spermatocytes, had therefore been reached. A female taken on 16 January had formed yolk. Thus, Mayer's careful anatomical observations have made it clear that in this area occupied bowers should be found from at least August onwards—in short, after the seasonal initiation of the sexual season. Display should be extremely active from September. The bowers will probably be deserted for a period following February. Shaw Mayer found fruit in the stomachs; insects were present in the ovulating female.

Weiske says that the nest of the builder of the above-described bower is found always in a pandanus palm at about the height of a man. It is an open, cup-shaped structure made of twigs 'and padded with foliage and shoots'. He never found more than one egg in a nest and this was yellowish-white. The call of the bird is a long-drawn shriek. The species is very shy and Weiske saw it only in pairs and then 'only when it was playing at the bower'.

Discussion

Although no hut is built, the central sapling or 'maypole' and the surrounding cylinder of moss are again the architectural basis of the bower. The decorated basal cylinder is homologous with the small cone built around the central pillar of the bower of *A. inornatus* (see Beccari's illustration, Pl. 20) and, likewise, with the decorated 'bank of moss' that is packed around the central pole which supports the hut-like shelter built by *A. subalaris*.

15

GOLD-MANED GARDENER
Amblyornis flavifrons, Rothschild

THE name 'Yellow-fronted Bower-bird', based on the specific name *flavifrons*, is not a good one. The bird is essentially brown-fronted in a manner not very different from *A. inornatus*. Further, several other bower-birds have, in one sex or the other, yellow or yellowish

FIG. 11. Male Gold-maned Gardener (*Amblyornis flavifrons*).
(*After Sharpe.*)

ventral parts. Likewise, 'golden-' or 'orange-crested' is unsuitable because of the similar crest colours of better-known members of the genus. The present bird is less than 9 inches in length yet it has a crest which is more than 3 inches long. This adornment glints in the light. It begins, not in the central crown as in two other members of the genus, but immediately behind the beak and flows back in a bright golden 'mane' of slender feathers which reaches more than half-way down the bird's back (Fig. 11). Therefore, in spite of the etymological impropriety of using such a word to describe an avian character, *A. flavifrons* will be referred to as above in the absence of a better term.

Below its spectacular golden-orange crest the bird's back is coloured dark brown. Its wings and tail are of a very similar hue. The throat is sooty-brown. Its chest is light brown, and the abdomen is light cinnamon-brown. The female remains undescribed.

The Gold-maned Gardener was collected in Dutch New Guinea at an unknown locality in the last century.[242] Nothing whatever is known of its behaviour.

16

QUEENSLAND GARDENER
Prionodura newtoniana, De Vis

IN September 1882 Kendall Broadbent,[30] a surveyor and a most discerning zoological collector as well, shot a drab-coloured juvenile female bower-bird in the remote Tully River rain-forest of north Queensland (Pl. 22). He sent it to the Queensland Museum where De Vis[67] described it as *Prionodura newtoniana* in honour of Alfred Newton of Cambridge, perhaps the most distinguished ornithologist of the day. Seven years later a Queensland Government Expedition was engaged in the biological survey of the wild uplands of the nearby Bellenden Ker region. On this expedition Meston shot a brilliant golden bower-bird at 4,800 feet. The date was February 1889 and the following month, 'while pursuing . . . official duties' in the Herberton rain-forests Broadbent met with 'several examples of a bird that I at once detected to be Newton's [= Broadbent's!] Bower-bird, and amongst them some gaily-coloured, full-plumaged cocks, which, instead of exhibiting the sombre hues of youth, are largely bright yellow-coloured . . .'. Broadbent wrote an official report of his rediscovery of the species on 30 March. Communications were by sea in those days, and before Broadbent's letter reached Brisbane, De Vis, with his usual celerity in such matters, had rushed a description of Meston's single male specimen into a local journal[69] and named his own *P. newtoniana* yet another genus and species, *Corymbicola mestoni*. Within a very short time this name had to be relegated to synonymy. Meanwhile, De Vis went farther and described the 'nest and eggs' of the new bower-bird as supplied by Meston. Campbell,[36] an egg-collector, declared that these probably belonged to a rain-forest 'fly-robin', but certainly not to a bower-bird. It was not until 1908 that the real nest and eggs of *Prionodura* were found.

The male of the Queensland Gardener has a small pale golden-yellow crest towards the back of the head. There is a patch of the same colour at the back of the neck. The whole of the ventral surface has this colour repeated. The tail, with the exception of the two inner feathers and the distal ends of the adjacent one, is pale yellow. Some birds are of a much deeper—almost an orange—hue than others. Close examination reveals that the yellow crest and nape exhibit a slight opalescent glint, not unlike that in the crest of *Chlamydera*, in certain lights. This play of colour is not obvious in the yellow plumage on the breast. The rest of the male plumage is brownish with an olive-green tint. The eyes are pale yellowish-white in both sexes. The beak is dark brown and the legs slate-blue. The female is olive-brown above with an under-surface of ashy-grey. The beak is like that of the male, but the legs are black. The juvenile plumage is not unlike that of the female. The bird is only 9½ inches long. The heads of both sexes are depicted in Fig. 3. The Queensland Gardener is the smallest of its kind in Australia, but it builds the largest of all bowers—relatively an enormous structure which sometimes reaches a height of 9 feet. It is, as we shall see, a maypole structure of the same group as those built by members of the genus *Amblyornis* in New Guinea (Fig. 2).

During the period of his rediscovery of the bird, Broadbent noted some of its habits. At the same time, other collectors were working in north-east Queensland, and it was soon established that *Prionodura* was undoubtedly a bower-bird, and descriptions of its gigantic bower were published. Very soon, too, the extreme restriction of its range became known (Fig. 19). It inhabits two pockets of dense mountainous rainforest. The bigger of these is the tangled sprawling mass of tropical forest which covers the Atherton Tableland and the Bellenden Ker Range at altitudes from about 1,500 to 5,000 feet. The second is the smaller tract which clothes Mount Cook, near the Endeavour River, some 125 miles northward. Here, near the summit (*c*. 1,470 feet), a male was collected by Olive in May 1899. It is not known whether the bird is plentiful on the relatively low Mt. Cook, but it is not uncommon in the Atherton uplands. Despite its brilliant plumage, the bird is inconspicuous in the forest and many timber-getters who have lived all their lives there are unaware of its existence. Like other bower-birds its principal diet is fruit, but it eats insects as well.

Bower and display

When a bower (Pl. 23) is first built it is merely a collection of twigs and sticks interwoven around the base of two saplings growing a yard or so apart. Each season the Queensland Gardener adds to its bower. The structures around the two saplings mount in height and are joined in between so that a thick wall-like building, roughly U-shaped, arises. In the centre, near the bottom of the 'U', a horizontal vine, stick, or root is left bare (Pl. 23). However high the bower becomes, this low perch remains clear. It is obviously a display-stick, perhaps broadly similar in function to the 'singing-stick' of the Stagemaker (*Scenopoeetes dentirostris*) (Chapter 21). The two pyramids of woven sticks become uneven in extent.[66] Day, who lived for nine years in north Queensland and collected 'several hundred' Queensland Gardeners, found a bower that had been added to until it became 9 feet high on one side and 6 feet 6 inches high on the other. The builder of this immense structure, it may be repeated, is less than 10 inches long. Most of the twin pyramids are roughly elongated oval at the base, but some of the bigger ones become nearly rounded. During the sexual season the inner wall of the larger pyramid and other parts of the bower fabric in the vicinity of the display-stick are decorated with pale moss, lichens, ferns, flowers, and odd bunches of berries. Jackson,[127] in November, found mosses that varied between yellowish-green and rich rust in coloration, and in addition open pods containing black and yellow seeds. This darker moss, as well as the empty pods, was seen by me in June, but it could not be determined what colour were the pods at the time they were placed on the bowers. The floral decorations generally consist of white-flowering epiphytic orchids (which sometimes continue to grow and flower in the fabric), in addition to other freshly plucked flowers. A few flowers are sometimes found on the ground in front of the bower.

The flowers are all placed upright inside their play-houses [wrote Day], but to see what the birds would do I once turned one of their orchids upside down. On the birds reassembling they made a great fuss and noise, and one of the old males replaced the flower in its proper position. I repeated the operation, and the flower was again placed upright by the old male.

Day claimed that the large bowers are 'resorted to by a large number of birds, and I have obtained over thirty at various times at a well-frequented play-house'. Day unfortunately gave little information concerning display, and his few important observations are undated.

Although these bowers are used by both sexes as a playground where they can chase and gambol with one another, they are frequently the scene of a pitched battle between a couple of adult males. This is caused by one male removing the flower placed in position by another, and a fight ensues, the remainder of the birds looking on and making a great noise, but not interfering with the combatants. I have never seen the females or young males fight; it is always the finest-plumaged old males.

Brief but important war-time observations were made by Bourke and Austin.[28] A drab-coloured bird was in possession of a bower early in October. This period, of course, is one of gametogenesis; and only a single bird was seen at the bower. On another occasion, another drab bird was seen at a large bower. 'A golden male appeared and swooped with a loud screech' at the drab bird 'which departed hurriedly, hotly pursued' by the golden male. As soon as the drab bird disappeared the golden male returned to the bower, hopped around for a few seconds, and departed.

'The drab bird then returned and played aimlessly around the bower for a few minutes, inspecting the flowers used as decorations.' The golden bird returned and drove the drab bird off once more. All this took place in less than 20 minutes, after which the watchers were compelled to leave. During their brief observations at various times, Bourke and Austin saw only one snatch of actual display. A male stood 'at the entrance of its bower', spread its golden tail, elevated 'the feathers on its nape' and bowed. This performance lasted for perhaps 10 seconds, after which the bird flew away.

Sub-bowers, too, are made by the Queensland Gardener. These are built near the principal bower. They are merely small structures that are similar to the beginnings of the large main bower (Fig. 2). Generally they are about 18 inches high. Broadbent found five within a space of about 10 feet and observed that they gave the spot exactly the same appearance as 'a miniature black's camp'. Five sub-bowers were observed near a principal structure by me in June. These were scattered over an area of 30 feet. In his writings Broadbent refers to the bowers of *Prionodura* as *gunyas* or *humpies* which are, of course, colloquial terms for native huts. This is interesting for it shows that although the structures of *Prionodura* are not hollow, the impression of an *Amblyornis*-like hut is obtained. Day found fifteen bowers of different sizes on Bartle Frere within a radius of a hundred yards but he does not give the relative numbers of large and sub-bowers. Sharp[257] said that only male birds assemble at the large bowers, and that the sub-bowers are the work of females. On several occasions Sharp removed moss, hung it some distance away, and then concealed himself to await the bird's return. On each occasion a radiantly coloured male bird, 'showing strong resentment', drew the moss back into position. Sharp conducted a second experiment suggested to him by aborigines: he set fire to three bowers. In each case a male bird flew up and perched near by, 'his beautiful head bowed and wings dropped, as though sorrowing over a funeral pyre'.

Like every other well-known bower-bird, *Prionodura* is a remarkable mimic. It has been heard to mock the Stagemaker, the Cat-bird, the Rifle Bird-of-Paradise, and other less distinctive birds as well. Whether it mimics at its bower has never been recorded. Near the bower it often uses a frog-like croak. Bourke and Austin say the 'wheezy croak' is preceded by a rattling noise like 'the rubbing together of branches'. The bird has also a single call which is said to resemble that of the Grey Thrush.

The display-season has never been exactly determined. Day found a decorated bower on the Upper Russell River in the middle of May, but bowers examined by me in the rain-forest between Wondecla and Atherton in June had moss, orchids, and seed-pods which probably belonged to the previous season's display. A male, however, was in the vicinity of each bower and, by means of a low whistle, it could be called right into a company position occupied by troops during training. Day found fresh flowers on a bower in early October on Bartle Frere. He further says that 'the birds do not frequent the bower during the wet season'. This generally begins late in November or in December, and is unleashed in full intensity from January to April.

Nesting season

The Queensland Gardener's nest and eggs were first discovered early in November 1908 on Bellenden Ker.[206] Nests are built also in December. The nest is a shallow cup-shaped structure about $5\frac{1}{2}$ inches in diameter with an egg-chamber $4\frac{1}{2}$ inches wide by $1\frac{1}{2}$ inches deep. It is made of dead leaves, bits of staghorn-ferns and dried mosses, and is lined with thin dry twigs. The nest first described was built about 3 feet from the ground in a split in a rotten tree. Most nests are built near the ground in similarly protected positions.

Two oval eggs are laid. Their lustrous unmarked white surface makes them utterly unlike the eggs of all other Australian true bower-birds, but similar to those described as belonging to species of *Amblyornis*.[280]

Discussion

The bower-form of the Australian *Prionodura* shows that it is related to the Gardener Bower-birds (*Amblyornis*) of New Guinea. It is the only mainland bower-bird which does not build an avenue, and, alone among the Australian forms, it lays white, unstreaked eggs.

Although the Australian bird builds a bower that may reach a height of 9 feet, the structure is demonstrably on the maypole plan as shown in Fig. 2. Its mode of construction enables it to be raised higher and higher in successive seasons so that ultimately gigantic (for a bird of its size) proportions are achieved. The precise function of the bower is unknown. The decorations so far recorded are almost exclusively pallid, although the bird has an almost unlimited range of colours at its disposal. It may be worth recalling that the eyes of the adult female, like those of the male, are pale yellowish white. Also, the most conspicuous part of the female when she is facing the bower—the chest—is also whitish in colour. There is no evidence, however, that in *Prionodura* these female characters influence the male's choice of colour-decorations; and there is certainly no apparent correlation between decorations and female colours in any of the New Guinea maypole-builders whose bowers have been described.

The few collectors' accounts of the display are, as usual, contradictory and of little worth. Some say that groups of birds of both sexes frequent the bower; another says that only the brilliantly coloured males use the principal bowers, and that the sub-bowers (see also the chapter on the Satin-bird) are built by the females. It was further stated that fights between males occur when one of them removes a flower placed in position by another. The above data may be brought into harmony if we assume that, at the period when the birds return to their bowers after the breeding season, a communal display may occur. Then, as the subsequent sexual cycle develops, the bower may become the property of the golden male owner who refuses to tolerate interference with his floral properties. Such a sequence of behaviour is familiar among the avenue-builders. The precise display to the female, which no doubt occurs at the bower, has never been described even though it has been stated that the birds can be observed at the bower without great difficulty.

It is clear that the sexual season is geared so that ovulation occurs at the beginning of the rainy season. Thus, when the young appear they will have at their disposal a superabundance of proteinous food.

The Queensland Gardener is certainly the only maypole-builder which still exists on the Australian mainland (Fig. 19, p. 174). Here it has been isolated in two restricted areas of tropical upland rain-forest by the xerophilous vegetation that has swept from the interior across the lowlands right through to the eastern littoral. Southward, the nearest extensive area of mountainous rain-forest is hundreds of miles away. To the northward, the nearest high rain-forest area is nearly 200 miles distant. Here the species does not occur. Leaving out of consideration this pocket of rain-forest, *Prionodura* is separated from the maypole-builders of New Guinea (*Amblyornis*) by about 350 miles of low-lying *Eucalyptus* and *Melaleuca* savannah and the waters of Torres Strait. *Prionodura* and *Amblyornis* are of common origin.

17

GOLD-CRESTED BLACK BOWER-BIRD
Archboldia papuensis, Rand

THE Black (Archbold's) Bower-bird was discovered in the Snow Mountains of Dutch New Guinea as recently as 1938 by members of the Archbold Expedition who did so much valuable work in the years immediately before the last war.[228] It is brownish-black, but its funereal appearance is relieved by two features: it has a small yellow patch at the margin of each wing, and the plumage of the upper- and under-surfaces alters in coloration in certain lights and takes on a 'scaled' appearance and a pale greyish sheen. The feathers have downy bases, with firm terminal portions which have 'in some lights a darker coloured tip'. This may be an epigamic feather adaptation of the same *genre* as the glistening lights in the plumage of the male *Ptilonorhychus*. It is significant that the pale sheen in *Archboldia* becomes less developed on the lower ventral surface which is, of course, not prominently exhibited to the opposite sex. The eyes of the above birds were dark, their feet grey, and the beak, which is unusually flat at the base, was black. The sexes were identical. The expedition secured a series of skins at between 6,000 and 9,000 feet.

Another skin, collected 10 years before those taken by the Archbold Expedition, was taken near Wisselmeren, in western Dutch New Guinea, at about 5,500 feet.[126] Further specimens have since been collected at 8,500 feet in the Mount Hagen uplands by Gilliard and Shaw Mayer. These appear to constitute at least a new sub-species and have been described as *A. p. sanfordi*.[190] The males of these central highland birds possess a prominent and unusual double-crest which is mostly golden in coloration (Fig. 12). Otherwise the plumage is jet black, completely lacking (in British Museum specimens) the yellow wing spot. Also, it lacks almost entirely the greyish sheen that enhances the feathers of the first birds described. The crest of the Hagen birds is erectile to a height of about $\frac{3}{4}$ inch near the beak. A second portion is continued backwards as a tract of feathers over the nape. Its golden-orange is mottled with black; the bases of the crest feathers are black and some of this can be seen among the gold. In a few feathers the gold is replaced by a lemon tint, and this shows occasionally amidst the black and gold. The Mount Hagen female is sooty brownish-black, with 'a vivid light patch' on the brownish wing. The throat is brownish-black and the concealed parts of the body are dark grey. It remains to be shown whether the original males discovered by the Archbold party were juvenile. If not, *A. p. sanfordi* may be a new species. Including the long, deeply forked tail the bird is about $13\frac{1}{2}$ inches long.

Shaw Mayer's notes reveal that the Black Bower-bird eats fruit as well as quantities of land-molluscs. The snails are carried to a special

FIG. 12. Male Gold-crested Black Bower-bird (*Archboldia papuensis*).
(*From skin in British Museum of Natural History.*)

stone or fallen tree-trunk where they are broken. A heap of broken shells marks the feeding-place. Mayer has again given gonad information which is of assistance in the determination of the breeding season. Late in March, testes measured only 4×2 and $1 \cdot 5 \times 1$ mm. On 22 April those of another bird were 6×3 and $4 \cdot 5 \times 3$ mm. in size. A male taken on 27 April had gonads measuring $8 \cdot 5 \times 4$ and 7×3 mm. From the above data it is probable that from April to June bower-display is energetic. No birds in moult, and with reduced testes, were shot, so the nesting period cannot yet be fixed with any certainty.

Rand[228] and Mayr and Gilliard[190] incline to the suggestion that *Archboldia* is related to *Amblyornis*. No details of the display-ground have yet been published. Two observers, however, have seen it, and I am indebted to these gentlemen, Messrs. Loke Wan Tho and E. T. Gilliard, for its description.

Loke Wan Tho says that:

the bower was a roughly cleared area on the ground which could not be compared with the beautiful finish of *Amblyornis*. The size of the playground is about 5 ft. by 4 ft. The trunk of a fallen tree and many branches above and about the playground were festooned with the bamboo-like trailing vine of a tree epiphyte. The floor was strewn with dead fern fronds. Some portions of the branches were quite bare, showing that they are frequently used by the bird to perch on. On the ground we found a pile of land snail-shells.

Gilliard says that the structure was 'nothing more than a collection of dried ferns, grasses, and sticks which had been trampled down in a kind of stage. Around its edges were several patches of snail-shells and one small pile of beetles' wings.' The display of *Archboldia* has not been described and neither have its nest and eggs.

Discussion

If the display-ground described above is full-sized and not rudimentary, it is still obviously a primitive affair that differs greatly from those of all other bower-birds. The bare branches above the display-ground and the bird's long tail suggest that an arboreal performance may take place. The rough description of a basal 'platform' of fern-fronds, grass, and sticks might at first sight suggest avenue-builder rather than maypole-builder affinities but only careful examination of a number of bowers and other display attributes will allow us to decide whether *Archboldia* belongs to one of the above groups or whether, more probably, it constitutes a distinct stem of its own. It is by no means unlikely that a second or even other members of such a third group may be discovered.

18

CRESTED BIRD-OF-PARADISE
(CRESTED 'GOLDEN-BIRD')

Cnemophilus macgregorii, De Vis

THIS beautiful bird is placed here, at the end of the list of true bowerbirds, because although it has been classified in the Ptilonorhynchidae, it is almost certainly a bird-of-paradise. A thorough anatomical investigation should show it is allied to the paradise-bird *Loria*, and perhaps also to *Loboparadisia*. Mayr[189] and Stresemann,[268] too, have grave doubts concerning its affinity with the bower-birds and believe that it may be a bird-of-paradise.

'The name *Cnemophilus* (Mountain-slope Lover) has been appropriated to it, and the species I propose, with permission, to dedicate to yourself.' So wrote De Vis[68] to Sir William Macgregor, the first Administrator of Papua, when he reported on the birds secured by the expedition, led by Macgregor himself, which explored the Owen Stanley Mountains behind Port Moresby in the year 1889. Macgregor must have earned fairly wide respect, for Goodwin, who also accompanied the party and who independently described two new birds, also proposed the name of the leader for the present bird:

The only new Bird of Paradise discovered during the expedition was a bird very similar to the Golden Bird of Paradise (*Xanthomelus aureus*) . . . It was met with on ascending Mount Owen Stanley. I therefore propose to name it *Xanthomelus macgregorii*, in honour of our leader—the first white man who has reached the summit of this range.

De Vis's description was published before that of Goodwin and so his name has priority. In any case, the new bird clearly is generically distinct from the true Golden-bird (Chapter 11). Its affinities, as we have seen, are still in doubt.

The male Crested 'Golden-bird' has a clear red head, merging into a golden-yellow back. The lower back and rump are brownish-yellow. The wings and tail are cinnamon. The feathered beak, sides of head to the level of the eyes, and the throat and the rest of the ventral surface, are rich chocolate-black. The whole of this is overlaid by an opalescent sheen that is apparent only in certain lights. The crest is peculiar: it is composed of several filamentous plumes, each about 1½ inches long, that are blackish-chestnut at the base and become deep orange at their tips (Fig. 13). The beak is brown. The legs are brownish-grey, and the eyes brown. The female has a reddish-brown upper-surface 'with a greenish shimmer'.[126] The throat and breast are somewhat paler. The female has also a small brown crest. The young male is said to be like the female. The bird is only about 8½ inches long.

The type specimen was secured at 11,000 feet. Since then many further specimens have been obtained at high altitudes. Blood collected

FIG. 13. Male Crested Bird-of-Paradise (*Cnemophilus macgregorii*).
(*After Sharpe, and from skins in British Museum of Natural History.*)

a series in the Hagen area and from these Iredale has described a new sub-species, *C. m. sanguineus*.[125] The male has a bright vermilion head and there are said to be other sub-specific differences as well.

The Archbold Expedition[229] found the bird fairly common, but very shy, in the Murray Pass region. Even the male, despite its brilliant upper surface, was very difficult to see in the shadowy lower reaches of the rain-forest. The bird feeds on fruit, and has a series of harsh, hissing, and clicking calls as well as a noise, 'startling in quality', that resembles 'the sound of two timbers being rubbed together under considerable stress'. No bower of the Crested 'Golden-bird' has been found. Neither, with any certainty, has its nest and eggs.[107a]

Discussion

Nobody has ever attempted to establish with any degree of confidence the precise relationship of *Cnemophilus* with any genus of bower-birds. De Vis suggested that the systematic place of the genus seems to be between *Xanthomelus* and *Amblyornis*. He[69] later described the bird-of-paradise *Loria* as 'a second species of *Cnemophilus*, a genus which, in deference to the opinion of eminent ornithologists, has been transferred provisionally to the birds-of-paradise'. However, De Vis soon came to the conclusion that *Loria* was different from *Cnemophilus* and proposed a new name for it. But the name *Loria loriae* was published by Salvadori[247] before De Vis's name was listed. In the past, *Loria* has been generally considered a bird-of-paradise and *Cnemophilus* a bowerbird. Iredale[126] says that *Cnemophilus* is a bower-bird but asserts that although at first sight it recalls the Golden-birds (*Xanthomelus* = *Sericulus*) it may not be closely related to them because of its unusual crest, its 'differently formed' tail, and its apparently dissimilar beak-structure. The bill, Iredale says, recalls that of the Australian *Prionodura*.

Here again the discovery of a bower, if the bird builds one, is of the greatest importance. It is extremely unlikely that a bower will be found. A recent American expedition[97] went to great lengths to ascertain whether *Cnemophilus* builds a bower. Hundreds of Melanesians were shown a skin and told that a fortune in trade pearl-shell and red calico awaited the man who could find its 'dance-ground'. The natives assured the white men that *Cnemophilus* does not build a bower nor does it display arboreally like other birds-of-paradise. It prefers the tree-tops where ' 'Im 'e walk about nothing tas all'—i.e. it undergoes no special display. If *Cnemophilus* is a bower-bird, so too, probably, is *Loria*. But it is highly probable that neither is closely allied to the Ptilonorhynchidae. However, if by any chance the bower does exist it should not be difficult to find. The meticulous recording methods of Shaw Mayer have enabled the display-season of the birds to be determined with a high degree of probability.

Mayer collected a series of *Cnemophilus* for the British Museum and sketched the gonad condition of the birds on the specimen labels. In the Mount Hagen area between 8,500 and 11,000 feet, two adult males taken late in June had testes measuring $9 \times 4 \cdot 5$ and $10 \times 4 \cdot 5$ mm. respectively. Thus, there are good indications that the sexual season was well under way and that, in fact, spermatozoa had probably appeared. Display should be pronounced at this period, but, at the same time, the actual reproduction date could be still many weeks ahead. In any case, the reproduction date of *Cnemophilus* can be determined by further data supplied on Mayer's labelled skins. He records that three adult males taken in November had testes which measured only 2×1 mm., $4 \times 1 \cdot 5$ mm., and $1 \cdot 5 \times 1$ mm., and that the birds were moulting. In the light of the information presented in Chapter 2, Mayer's data can be interpreted only as follows: The seasonal display is well established in June and full spermatogenesis occurs between July and September. Reproduction takes place before the middle of November. The moult mentioned by Mayer is the characteristic post-nuptial moult that accompanies the seasonal metamorphosis of the testes after the shedding of spermatozoa. This conclusion is reinforced by a female, collected in November, which contained only minute oocytes and was also in moult. If *Cnemophilus* is, in fact, a true bower-bird its bower can be found in the Mount Hagen area during the period from June to October.

Mayer, incidentally, recorded also the testis condition of a juvenile which had become almost completely dark chocolate on the ventral surface with the exception of some yellow mottling on the abdomen. Red had appeared at the base of its crest. One testis measured 6×2.5 mm. and the other 1.5×1 mm. This larger testis must have contained spermatocytes. It will be recalled that male Satin Bower-birds undergo complete spermatogenesis and reproduce while still lacking the scintillating plumage of the adult.

19

GREEN CAT-BIRD

Ailuroedus crassirostris (Paykull)

THE Cat-birds, although they are almost universally considered to be members of the Ptilonorhynchidae or bower-birds, do not themselves, as far as is known, build bowers. Gould[104] suggested that the Green Cat-bird (Pl. 24) 'might construct a bower similar to that of the Satin Bower-bird'. He was unable to satisfy himself that it did. It now seems certain that neither of the two Australian races of the cat-bird builds a bower, but from time to time, until today, Gould's surmise has led naturalists to suggest that they occasionally do so.

The Green Cat-bird takes its name from its most beautiful, almost emerald, green wings and tail and its voice of which Gould wrote:

Its loud, harsh and extraordinary note . . . differs so much from that of all other birds, that having been once heard it can never be mistaken. In comparing it to the nightly concerts of the domestic cat, I conceive that I am conveying to my readers a more perfect idea of the note of this species than could be given by pages of description. This concert is performed either by a pair or by several individuals, and nothing more is required than for the hearer to shut his eyes to the neighbouring foliage to fancy himself surrounded by London grimalkins of house-top celebrity.

Although the colonists, before Gould's day, named it the 'Cat-bird' many later observers have not agreed that the calls resemble those of a cat as closely as is often suggested. Some have likened the notes to the crying of a child, and the call of the northern *A. c. maculosus*, an interesting example of regional variation in bird-song, is still less like the mewing of a cat. However, there can be no doubt that the cry, 'having once been heard . . . can never be mistaken'. In addition to the cat-like cries, the species has also a single, thin, alarm-call which is of a surprisingly limited volume for so strong-voiced a bird.

In addition to its bright green dorsal surface, the Green Cat-bird has a yellowish-green breast, mottled paler. Its green wings have pale spots. The head and throat are olive-brown, mottled whitish. The beak is horn-coloured. The eyes are reddish-brown and the feet are brown. The sexes are externally identical and are about $12\frac{1}{2}$ inches long.

Details of the discovery of the Green Cat-bird are unknown. Somehow, a skin reached Sweden in the early years of the last century, where it was described as a shrike, *Lanius crassirostris*, by Paykull in a journal from Uppsala.[215] This description was overlooked for many years. Meanwhile the bird was described as a *Coracina* (or 'cuckoo-shrike', as some people call it) by the Frenchman Vieillot. Next it was classified as a bower-bird, *Ptilonorhynchus smithii*, by Vigors and Horsfield. In 1851 Cabanis saw that it was quite unlike the Satin-bird and gave the

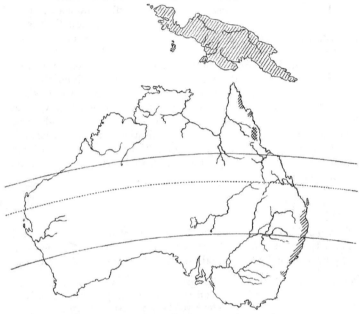

FIG. 14. Approximate distribution of the Cat-birds.
Stripes: Genus *Ailuroedus*.
Cross-hatching: Genera *Scenopoeetes* and *Ailuroedus*.

bird its present generic name. Later, when Paykull's earlier description became known, the specific name *smithii* of Vigors and Horsfield had to be replaced by the earlier *crassirostris*. So it is today.

Alfred Russell Wallace, travelling in the New Guinea region, collected in the Aru Islands a bird that was obviously closely related to the Australian Cat-bird but which was given specific rank (*A. melanotis*). This bird has a more blackish-brown head and throat, and is more prominently spotted. Later it was shown that the Aru Islands bird is widely distributed through the lowlands and low ranges of southern New Guinea and, further, that it occurs in north Queensland as well (Fig. 14). Meanwhile, however, it had been described from Queensland as *A. maculosus*. This bird has been relegated, with what appears to be sufficient reason, to subspecific rank.[191] This leaves us with the nominate race *A. c. crassirostris* ranging through the coastal rain-forests from southern Queensland (Blackall Range) to southern New South Wales (Shoalhaven River), and a northern race, *A. c. maculosus*, living in the restricted Cairns/Atherton Tableland/Cardwell area of northern Queensland. In addition to *A. c. melanotis*, several further sub-species have been described from both high and low altitudes in New Guinea.[188] *A. c. crassirostris* and *A. c. maculosus*, incidentally, have mated and reproduced in captivity.[126]

Even though the Green Cat-bird is usually not uncommon throughout its range, comparatively little is known of its behaviour. It is not shy and, if the observer remains still, it will often come quietly to within a few feet to investigate. Its food is chiefly fruit, but although it will eat domestic fruits it does not go far from the rain-forests, with which its plumage harmonizes perfectly, to raid cultivations. In the rain-forest it often feeds in company with Satin-birds and Regent-birds, and in north Queensland it eats land-snails. Both northern and southern forms are known to eat insects.

Display

Gwynne[106] has given a valuable account of what appears to be a sexual display. In October, in the Lilyvale area south of Sydney, he found a bird pursuing another through the open timber fringing the rain-forest.

During this flight both of the birds gave the usual drawn-out, cat-like calls, one chasing the other in and out of the trees. Finally they perched, and after a short period began calling again for several seconds. After the calls they spent about twenty seconds preening their feathers, then the chasing operations were resumed, both acting as before.

There was a lull in this behaviour, but it was continued later. Gwynne found an almost-completed nest in a heavily forested gully nearby.

The cat-like cries are sometimes heard in the earliest hours, and at almost any time during daylight. In the first light of dawn and a little later there occurs another, and most unusual, form of vocal display in the southern race of cat-bird. This is given by an individual perched on a vantage-point low in the rain-forest. The sound sequence begins with one or more sharp clicking notes, followed by three very loud guttural cries. Two of these are long-drawn; the third is brief and clipped. This sequence may be repeated for some minutes but it seems rarely to be given after sunrise. It has been so far heard in the middle of June, late in September,[179] and in the middle of October.[86]

Nesting season

The Green Cat-bird ovulates in October, usually a little before the Satin-bird in areas where they coexist. It generally builds amid a tangle of vines, but occasionally it chooses a less sheltered situation such as the top of a tree-fern. Most nests are between 10 and 20 feet from the ground, but both northern and southern birds may build somewhat higher or lower. Although usually well-concealed, the bulk of the nest often makes it fairly easy to find. It is a substantial, open, cup-shaped structure composed of vines and twigs with which are interwoven numerous broad leaves. It is lined with thin vines and fine roots. It is 8 to 10 inches wide by 6 to 8 inches deep. The egg-chamber is about 5 inches wide and 3 inches deep (Pl. 24). It is therefore a much heavier structure than those built by most bower-birds. There is an unverified statement[263] that *A. c. maculosus* 'decorates its nest'.

Two oval, cream-white eggs are generally laid but occasionally only one or as many as three may be produced. In New South Wales the Green Cat-bird hatches its young generally in November and December. In northern Queensland the breeding time is much the same. Here the young are launched at the beginning of the rainy season which, of course, ensures an adequate protein harvest for their development. Miller[195] says that the young chicks are fed entirely on insects, while young birds about to leave the nest, when perhaps 3 weeks old, are given an essentially vegetarian diet.

Rand[229] found *A. c. melanotis* nesting in New Guinea from September to December. The eggs were light olive-brown in coloration. He says that both parents fed the young. In 2 hours 10 minutes the male brought food on five occasions. The female came four times but brooded for awhile during each visit. Rand says that the young were 'evidently fed on fruit'. Here we see a most profound departure from the behaviour of the true bower-birds, for, as far as is known, no male bower-builder ever approaches the nest.*

The Green Cat-bird bands into small foraging flocks during the autumn and winter months, but these rarely move far afield or contain more than 15 or 20 individuals.

* See also Addenda, note 'E', p. 192.

20

WHITE-THROATED CAT-BIRD

Ailuroedus buccoides (Temminck)

THE handsome White-throated Cat-bird (Fig. 15) is confined to New Guinea and adjacent islands. Here it is extremely widespread, ranging from the islands off the north-western Vogelkop, such as Waigeu, Salawati, and Batanta, throughout many parts of the New Guinea mainland to the far south-east. The species was described in

FIG. 15. Male White-throated Cat-bird (*Ailuroedus b. buccoides*).

(From skin in British Museum of Natural History.)

1835 as *Kitta buccoides*[272] from a specimen taken in the Triton Bay region of Dutch New Guinea, but it penetrates far from the coast into the upper Fly River and elsewhere.[188] Rand[229] says that the altitude range of the species is greater than that of *A. crassirostris*. Its precise distribution is unknown.

The White-throated Cat-bird is usually known as the 'White-eared' Cat-bird, but as the white on the throat is more conspicuous than that near the ears I believe that the present name is the better. The bird has a dark olive-brown head, a white throat, white ear-patches, and a breast that is rich buff-colour, heavily spotted with black. The abdomen is unspotted. The back and tail are bright green. The wings are bright green and brown. The eyes are brown, the bill horn-coloured, and the legs bluish-grey. Young birds have blackish beaks and darker crowns. The sexes are externally very alike. The bird is about 9½ inches long and is therefore considerably smaller than *A. crassirostris*. It varies a great deal: its races are listed by Mayr.[188]

Apart from a few remarks concerning its calls, very little of its behaviour has been recorded. One collector likened its hissing alarm-note to the 'spit of an angry cat'. Weiske[280] says it lays yellowish-white eggs. A 'laying female' has been collected in May.[229]

21

STAGEMAKER

Scenopoeetes dentirostris (Ramsay)

THE Tooth-billed Cat-bird, 'Bower-bird', or Stagemaker, is restricted to the wet mountainous forest lying approximately between 16° 58′ and 19° 0′ S. a few miles inland from the coast of tropical north-eastern Queensland (Fig. 14). Its range is much the same as that of the Queensland Gardener Bower-bird except that it has not yet been recorded from the Cooktown pocket of rain-forest a little farther north. Its southern limit is the Mount Spec rain-forest on the Seaview Range about 40 miles north of Townsville.

The Stagemaker has been variously considered both a cat-bird and a bower-bird. It is clearly a modified cat-bird. It has taken partly to the rain-forest floor, and has replaced the obliterative bright green plumage of the cat-birds with an equally cryptic brownish colour-scheme that harmonizes beautifully with the fallen leaves and dark earth of the shadowy forest. Here it builds its curious leaf-strewn 'stage' (Pl. 25) from which Ramsay[224] gave the bird its generic name *Scenopoeus* (Greek: 'Stagemaker'). Coues,[58] however, pointed out that Agassiz had pre-occupied this name with an insect, and suggested the present one. Its specific name, *dentirostris*, comes from its 'toothed' beak (Pl. 25) which is an adaptation related solely to the habit of sawing leaf petioles in two, preparatory to the daily decoration of its display-ground.

This beak has three pronounced depressions in the lower mandible and there are corresponding serrations, not visible outside, within the upper mandible. Both sexes have the 'toothed' beak and they are externally similar in all other respects as well. The dorsal surface of the bird is brown. Below, it is fawn, mottled with brown. It has reddish-brown eyes, a beak and legs of much the same colour, and it is about $10\frac{1}{2}$ inches long. It lives essentially on fruit, but it eats insects and large ground molluscs which it sometimes cracks on a special stone that is often near the display-ground (Pl. 25). On the rain-forest floor near, but not on, the display-ground, a small pyramid of voided fruit-stones is usually found. Jackson[127-8] says that after digesting the fruit, the Stagemaker probably regurgitates the stones from its 'singing-stick' on to the ground. He found *Nephelium* seeds on the pyramid. Several sorts of seeds were found by me on pyramids in 1944. Two were identified.[84]

One, a long seed with an aromatic odour, belonged to the tree *Cryptocarya Mackinnoni*. The second, a globular, ribbed seed, was from *C. obovata*. Seeds taken from the intestines of *Scenopoeetes* were again of several species. One was identified as belonging to a species of *Elaeocarpus*.

The Stagemaker was discovered by Johnstone in 1874 on the Seaview Range and was for many years thought to be a rare bird. This was merely because its habitat, in those days, was very remote from civilization. Even today the wild mountain forests where it lives are sparsely populated, and so very few living naturalists have seen it. The same is the case with *Prionodura*. However, the Stagemaker is far more abundant than the Queensland Gardener and, in fact, is one of the commonest species in parts of its area. Further, despite its perfectly obliterative plumage, it is, during the sexual season, an easy bird to find and to collect. Its habit of perching on a special 'singing-stick'—a twig or liana —above the display-ground, and sending an almost constant stream of song into the forest, quickly leads the collector to its territory.

Display season and associated phenomena

Display-grounds have been found only from August to December inclusive. Testes collected by me in the middle of August, before the construction of the stages, had narrow tubules which contained only spermatogonia (Pl. 26). Late in August birds were often heard calling in the canopy and at about this period the first display-grounds were discovered. Early in the season the birds are very reluctant to approach their stages if they have evidence that they are observed. At this period one may wait by a stage for hours without seeing its owner. Light is now increasing in duration at a rate of about 40 seconds per day. It is difficult to believe that this fluctuation could stimulate the reproductive rhythm to seasonal activity. There is in the rain-forest an annual temperature deviation of only about 13° F.; but it is during August and September that the seasonal increase in temperature begins. Between August and November there is a rise of more than 10° F. From August to October there is also an appreciable drop in humidity. This is also the driest time of the year. The seasonal sexual resurgence, as proved by gonad condition, begins in August and September.

As the days pass, more and more display-grounds are built and the volume of song increases sharply from above them. By September, foliage that is often used by the birds is sometimes seen with numerous petioles chewed and 'sawn', with varying degrees of success, by Stagemakers which have been trying to free the leaves for their display. Leaf-gathering has been described.[128, 170] The bird employs a sawing or grinding motion until most of the stalk is severed and the leaf can be tugged free. The bird must work hard to sever a single petiole of (for example) *Alpinia racemigera*, which is about 5 mm. thick. This species, the wild ginger shrub, is favoured by some individual birds and neglected by others even though it appears to occur throughout the Stagemaker's range. After some 12 minutes' intermittent labour on one petiole, a bird I saw then turned to another which had been attacked earlier. After some minutes' further grinding and tugging with its beak, it succeeded in freeing the leaf which was then carried off towards a display-ground some 40 yards away. In the case of a petiole that could not be attacked from the ground, the bird flew a couple of feet into the air and bore the leaf to the ground where sawing operations were begun. It is of outstanding interest that a sharp downward tug of the human hand will instantly disarticulate many of the leaves used by *Scenopoeetes*. Yet an examination of the petioles of nine kinds of leaves on many playgrounds indicated that most, and probably all, had been parted by the laborious sawing technique. It would seem, therefore, that there has been developed an elaborate beak-structure that performs arduously a task that could be done far more simply if the avian brain could discover how. The bird apparently has considerable difficulty in severing some stems which are only 2 mm. thick. On one display-ground, along with thirty-two other leaves of four species, was found a 10-inch *Sideroxylum* twig of 8 leaves. Examination of the thick twig showed that it had been sliced from the tree by the bush-knife of a passing soldier. A bird had attacked each of the eight petioles but had not persevered sufficiently to sever any one of them. Finally it had taken the entire complement to its display-ground. Each leaf was carefully arranged (with the lighter undersurface uppermost) so that it would not touch the others on the twig.

Jackson took particular note of the foliage on the stages investigated by him in 1908 and found that leaves from fourteen species of trees or shrubs were used. The three commonest were given as *Cryptocarya mackinnoniana*, *Litsea dealbata*, and *Tarrieta argyrodendron*. In 1944 I examined dozens of stages.[170] The leaves on what seemed to be five typical ones were collected and samples of each species were sent for identification to the Botanic Museum, Brisbane.[84] The five stages contained leaves of eight species of trees or shrubs. The varieties of leaves, their approximate sizes, the thickness of the petioles, and the size of each stage and its locality and date are given in Table IV.

A total of at least seventeen kinds of leaves have so far been found on stages. The leaves of *Litsea dealbata*, which figured so prominently in Jackson's lists early in the century, were absent on the stages tabulated on p. 157. There was no doubt about this because the botanist who identified the plants was asked particularly to look for *Litsea* in the samples collected. It is unlikely that Jackson's identifications were incorrect.

Although insufficient specimens were collected for concrete conclusions to be drawn, there is some evidence that certain birds have a predilection for certain kinds of leaves. When one-third of the leaves (numbering 12, 13, and 15 respectively) were experimentally removed from three stages in November, and replaced with like numbers of leaves from other

TABLE IV

Size, location, and date of stage	Species of leaves on stage	No. of leaves	Approx. size of leaves	Approx. thickness of petioles
1. (5½′ × 4′) Cassowary Track. 27 Oct.	Alpina racemigera (wild ginger)	15	20″ × 7″ to 12″ × 4″	5 to 6 mm.
2. (8′ × 8′) Cassowary Track. 26 Nov.	Darlingia spectatissima (a 'silky oak')	13	15″ × 9″	2·5 mm.
	Cryptocarya (sp.?)	19	14″ × 8″	5 mm.
	Sideroxylum (nov. sp.?)	8	5″ × 3″	2 mm.
	Alphitonia Whitei (a 'red ash')	1	7″ × 4″	1·5 mm.
3. (5′ × 4′) Lake Eacham. 6 Dec.	Tarrietia Argyrodendron (var.) (booyang, or 'crow's-foot elm')	28	6″ × 3·5″	3 mm.
	Croton (sp.?)	14	7″ × 2·5″	3 mm.
	A. racemigera	9	12″ × 4″ to 20″ × 7″	5 to 6 mm.
4. (5′ × 4′) Tower Hill Reserve, Atherton. 12 Dec.	A. racemigera	36	20″ × 7″ to 12″ × 4″	5 to 6 mm.
	T. Argyrodendron	6	6″ × 3·5″	3 mm.
5. (10′ × 3′) Tower Hill Reserve, Atherton, 12 Dec.	Legnephora Moorei	32	6″ × 4″	3 mm.

plants, all the alien leaves were removed by the following morning and partly replaced with leaves of the species originally used on the stages. Also, it seemed that some birds would pass a species of plant (e.g. wild ginger) favoured by others in order to get leaves of a sort that they personally preferred. It was unfortunate that duties in an infantry battalion, camped not in the rain-forest but in savannah several miles away, made impossible anything but brief spasmodic experimentation.

The birds did not place leaves in precisely the same positions each morning. The principal desire seemed to be the attainment of a striking pattern of leaves, however haphazardly they were arranged. When I turned all the leaves over, the bird rearranged them to its former preference, even though it is not very unusual to find a stage with an odd leaf with darker upper side uppermost.

Jackson claimed that the birds display 'foresight' in putting the leaves upside down because, so placed, they curl less rapidly in the tropical heat. It is more probable that the bird places them with paler surfaces uppermost solely to achieve a more spectacular visual contrast with the dark floor of the rain-forest.

Display-grounds are generally roughly oval or circular and measure between 3 and 8 feet in diameter. Proportions may vary in relation to nearby obstructions such as tree trunks. One stage measured 8×2 ft. and another 10×7 ft. The latter, quite apart from its large area, was of unusual interest in that half of the oval, though meticulously clean, was completely bare of leaves whilst the other half contained no less than 40, closely packed. This seems to be about the average number of leaves on each display-ground, but Cornwall[56] found a stage containing 75 and Campbell and Barnard[37] record one with 103. Jackson examined a stage decorated with leaves of 13 species, but most display-grounds contain only three or four kinds. Sometimes stages are decked with leaves of one species only. Some leaves, notably those of *A. racemigera* and *D. spectatissima*, may be about twice as long as the bird that carries them.

As the leaves wither they are replaced by fresh ones that appear generally to be gathered in the early morning. The discarded leaves are abandoned a foot or so from the edge of the stage. One writer described a ring of dead leaves over 1 foot high around one stage. A bird has been seen to 'fan' a drying leaf off a stage with its wings, but it is not known whether this practice is general.[28]

There is as yet no good evidence of any prolonged physical display by the bird on its stage, though Jackson saw birds descend from their singing-sticks, 'flirt with a leaf and fly back again to their old post and old performance' (of mimicry). Bourke and Austin described a little 'strutting' with lowered wings. The stage-owner's physical display, as far as is at present known, seems to consist principally of loud vocal advertisement from a chosen 'singing-stick' from 2 to 10 feet above the display-ground. From this special perch—a twig or liana—the bird sends out a more or less constant medley of sound through the thick rain-forest. This noise, a mixture of its own varied notes and the faithfully mimicked calls of neighbouring species, can be heard at a considerable distance, although the mottled and almost motionless singer is not readily visible at more than a few yards. Stages are often found within 100 yards of each other.[170, 262] Jackson mentions a couple only 36 yards apart, but this is exceptional. In some dense and hilly areas, the voices of five or six birds can be heard simultaneously by the human ear. The owners of the stages appear to answer each other. Some call almost constantly for more than an hour on end. There is also a low musical soliloquy that can be heard only a few yards away.

Thus, although the sombre Stagemaker does not indulge in the spectacular posturing of the more brilliantly coloured males of the true bower-birds, it nevertheless draws constant attention to its presence, and to the whereabouts of its hidden display of leaves, by a constant long-range vocal advertisement. Associated with this continuity of sound is a notable, though not very obvious, morphological adaptation (Fig. 16). When a bird sits silent in the shadowy sub-canopy its cryptic coloration makes it almost invisible. But when it begins calling, the buccal movement causes a motion among the feathers below the beak and this leads to a slight reorganization of the colour pattern. Then the yellowish bases of certain feathers become visible and, at the same time, paler parts of the palate are revealed. The precise situation of the singer is now easier to locate. When the bird ceases to sing it is once more almost invisible against its shadowy background. It is likewise difficult to see when approaching or leaving its stage.

Almost all travellers and collectors have stressed the singularly penetrating quality of the voice of *Scenopoeetes*. Lumholtz[154] was drawn to its 'place of amusement' by its 'loud and unceasing voice'. Again, all who have remained long enough in the rain-forest to become familiar with the calls of neighbouring species have commented on the fidelity with which the Stagemaker mocks so many of them. Broadbent considered it second only to the Lyrebird as a mimic. Altogether, it has been listed as imitating the calls of at least twenty-six other birds as well as the distress calls of frogs and cicadas when taken by predators. The notes of other birds are so woven into the display-song that, until the observer becomes very familiar with the voices of other species of the area, he has no clear idea which notes are 'borrowed' and which are part of the caller's own varied repertoire. The sum total of melodious noise, however, certainly ensures that every rival or potential mate within a wide area is aware of the focal point of the caller's territory.

There is as yet no reliable evidence that more than one bird regularly attends each stage. A nineteenth-century author[79] stated that the Stagemaker 'makes clear spaces where from six to eight males meet to sport ... decorating them with green or coloured leaves, berries and flowers'. Chisholm[46] has quoted collectors' information that the number of leaves used depends upon the numbers of birds in attendance at the display-ground upon which they 'frisk and toy [with the leaves] in obvious enjoyment'. North,[205] too, has nineteenth-century collectors' information that three or four birds have been seen turning the leaves and playing

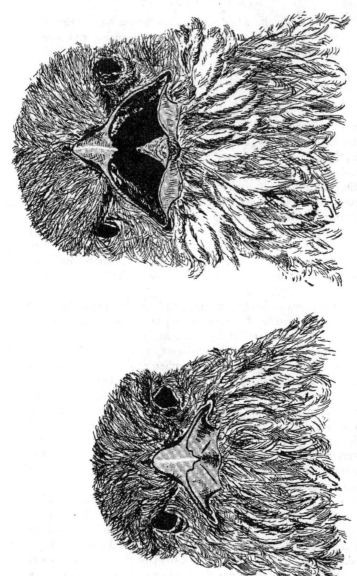

FIG. 16. The Stagemaker (*Scenopoeetes dentirostris*), showing reorganization of colour-pattern during song.
Left: Silent, with beak closed.
Right: When the bird sings the buccal movement causes the yellowish bases of the throat feathers to be revealed and the bird becomes easier to see in the gloomy rain-forest (del. C. E. Pugh).

with them, while another bird perched nearby 'pouring out its loud melodious notes'. North cites another account of half a century ago which suggests that from two to a dozen birds frequent a stage, tossing its leaves about. These statements have not been verified. The reference to coloured leaves, berries, and flowers on the display-ground was probably the result of nineteenth-century naturalists confusing accounts of this species with those of the true bower-birds. Although apparently without foundation, this information has been copied from time to time ever since.

By the middle of September the birds are calling loudly from above numerous stages and show little hesitation in returning to the singing-sticks provided the observer remains fairly inconspicuous and not immediately adjacent to the leaves. During this season song and mimicry occur also in trees some distance from the stages. I removed singing birds from above stages and a week later the singing-sticks were occupied by other males in similar gonad condition, and freshly gathered leaves covered the ground below. Fresh stages were prepared as late as October. Sometimes a singing bird attracts attention to a cleared space that is as yet undecorated. November is the height of the display season. The seminiferous tubules of testes examined during this month contained small bunches of spermatozoa, but few sperms were as yet free in the lumina. The interstitium was considerably dispersed and individual Leydig cells were of maximum size (Pl. 26). During November some forty stages were examined briefly. All were decorated, though late in this month some of the drying leaves were not replaced with the promptitude shown earlier in the season. This situation continued through the first week in December, but now more and more stages were partially neglected. Odd dry leaves remained unchanged. By the middle of December more than half the display-grounds had fallen into partial or even complete neglect. Song and mimicry on the singing-sticks now waned in intensity, though many birds still called loudly. Testis tubules were at a maximum expansion and contained thick bunches of spermatozoa; sperms were free in the lumina and the first signs of tubule metamorphosis had appeared. Interstitial cells were dispersed throughout the greatly enlarged organs (Pl. 26).

Jackson, too, is emphatic that mimicry increased in intensity as the display-season advanced and that late in December, when the birds neglected their display-grounds, a 'comparative silence' fell upon the forest. This contrasted with the 'babble of mimicry' heard during the previous six months.

Nesting season

The Stagemaker makes its nest in November and in December. This is a flimsy, saucer-shaped structure built from 17 to 100 feet aloft in the rain-forest and often extremely hard to find. It is made of twigs and is about 5 inches in diameter and 2 in depth. Two eggs are laid. These range from cream to creamy-brown and are oval.

The earliest young on record were found on 10 December. November and December, the months of nidification, signal the beginning of the wet-season. In December, in normal years, the deluge becomes typically monsoonal. From February to March rainfall remains extremely high (averaging about 17 inches per month) and in April there is a sharp drop. This dry trend usually continues until the following October. The last-named month is the driest of the year, with an average rainfall of less than 3 inches. The Stagemaker, then, brings out its young during the wettest period of the year. It is at this time, of course, that insect food is most abundant. Jackson found that the young were fed on beetles even though, as we know, the adults are largely frugivorous.

It is not yet known whether the species has a post-nuptial flocking phase. It has been claimed[46] that display-grounds are 'resorted to all the year round' but there seems to be as yet no admissible evidence that they are used in any period outside that of August[105, 170] to January. (The observations of Cornwall[55] in the late days of December seem to make it certain that some stages are still in use early in January.) Coleman[52] asserts that in February all stages are deserted, but that the birds are still present in the canopy. Jackson says that during April the young (hatched in December and January) are still being fed by their parents, but that there are no display-grounds in the forest. There have been suggestions that the birds leave the breeding forest after the young fly. Green[105] states that they disappear from February until late in July or early in August. Coleman says that the birds are 'never much in evidence' between February and August. Whether the Stagemaker has a post-nuptial nomadic phase, or whether it merely remains silent and inconspicuous in the canopy until the following display-season is not known. There appears to be only one record of the bird outside the mountainous nuptial area: North[205] says that the collector Grant shot a pair near the coast 'opposite Double Island'. No date was given. Lumholtz said he found it only 'on the summit of the coast mountains in the large scrubs [= rain-forests] which it never abandons'.

In one locality there is evidence that a limited colonization is going on. This is near Atherton where a conifer plantation has been established. At first this contained no undergrowth but during the war years a low growth of lawyer-vine and other shrubbery was allowed to establish itself. The plantation was invaded by *Scenopoeetes* which became very common there even though visibility in this artificial area was far greater, and the birds' concealment less effective, than in its natural rain-forest environment.

Discussion

The sexual and display cycles of *Scenopoeetes* run parallel. Whilst it cannot be denied that the birds get aesthetic pleasure from their stage-decoration and from the song that they produce from above it, these activities, extraordinary as they are, form essentially a highly specialized mechanism that ensures the acquisition and establishment of territory, the attraction and retention of a mate in dense rain-forest, and the synchronization of the male and female reproductive processes before the female can be inseminated.

We have as yet no clear idea of the more precise functions of the display-ground. It is certain that the almost constant babble of noise emitted by the bird enables all rivals and potential mates over a wide area to know where the caller and his stage are located. The inclusion of the calls of other birds is one of the most remarkable features of the vocal advertisement, but the phenomenon is common to many other rain-forest birds in Australia. I have shown[167] elsewhere that the widespread belief that there are more vocal mimics in Australia than anywhere else is fallacious, but it is true that that country has a much greater number of *expert and constant* mimics than has been recorded from any other region. These notable mimics almost all dwell on or near the ground in dense rain-forests or in other areas where visibility is more or less restricted. Unable to draw attention to themselves by such devices as bold plumage pattern, and song from vantage-points visible from afar, rain-forest birds (like nocturnal birds) have developed loud and penetrating voices and are thus able to keep in touch with each other. Some thicket-dwelling species have developed a sustained song of their own, but many, especially the builders of display-grounds, have not done this and achieve a sustained effect by mixing their own loud calls with those of other birds. Mimicry *per se* has biological value in that the young of many species must learn their song from others of their kind.[122] Nearly 200 years ago it was discovered that young sparrows, put with adult Nightingales, learn to sing rather like Nightingales.[20] The emergence of special powers of mimicry in the selective medium of the rain-forest floor is discussed in the following chapter.

All nine examples of *Scenopoeetes* secured on singing-sticks by myself and helpers were males.[170] Lumholtz refers in the masculine gender to a display-ground caller that he described, so it is probable that he killed a male on its singing-stick. Johnstone's type specimen, described by Ramsay, was shot with a rifle and was too battered for the sex to be determined. Jackson, on the other hand, asserts that he collected two females singing above stages. Certainly both sexes possess the leaf-cutting beak.

We have only one brief indication of the possible precise part played by the leaf-covered stage during courtship. A dead bird (which on dissection proved to be a male) that had been taken from above one display-ground was placed on a branch a few yards from the singing-stick of another. Within a few seconds the owner of the second stage appeared. Instead of assailing the dead male violently (as the brilliantly coloured male of the dimorphic Satin-bird does in similar experiments), the owner of the stage flew down to the edge of its leaf-display. It partly concealed itself from the watcher and, from behind a tree-trunk, called with a note strongly reminiscent of that which the Satin-bird employs when its mate leaves the bower and it wishes it to return. It seemed that the owner of the stage desired the dead bird (the sexes are identical in external appearance) to come from the low branch towards it and the display-ground. The bird was unfortunately startled by passing troops and flew away. The date was 26 November and the testes of both birds contained bunched spermatozoa.

Although the display-ground of the Stagemaker has some of the functions of the bowers of the true bower-birds it is utterly different, both in architecture and decoration, from either of the two larger groups of bower-birds and it is not very like the display-ground of *Archboldia*. The relatively simple arena of *Scenopoeetes* does not appear to be a stage in the evolution of display-ground building in the Ptilonorhynchidae. Rather, it is the culmination of a development that has proceeded independently within the cat-bird family. Any similarity in its functions with those of true bowers appears to be the result of convergent evolution.

22

THE EVOLUTION OF BOWER-BUILDING

I BELIEVE that the display habits of the bower-birds, however bizarre they appear on casual consideration, are in fact no more than the extraordinary elaboration of the territorial and display impulses that are found in other quite commonplace birds. Avian display—visual, auditory, or a combination of both—is usually associated with conflict and the establishment of territorial domination (including often the spacing out of the population), the formation of the pair-bond, and the development and synchronization of the sexual processes of the pair until the environment becomes seasonally appropriate for the female to build her nest and to rear her brood. Not before will she accept the male, however ardently and energetically he postures beforehand. It is the same with the bower-birds in so far as they are known.

We can see all sorts of intergradation between the simpler forms of display of other species and the elaborate colour-selection, rhythmically noisy display 'dances', and display-ground architecture of bower-birds. Numerous species, as widely diverse as pheasants, grouse, waders, lyre-birds, manakins, and birds-of-paradise, make simple display-grounds and posture on them during their sexual season. These display-grounds are examples of convergent evolution—as is also the development of loud, penetrating voices and high powers of vocal mimicry among species of quite dissimilar passerine families which live in rain-forests where long-range visual display is impossible because of restricted visibility. Convergent evolution—and adaptive radiation—are as true of display as they are of physical characters.

The bower-birds, and one cat-bird, differ from the many birds which construct a simple display-ground in that they habitually bring to this, the focal point of their territory, properties that they use to enhance their performance. The bower-birds have developed still further—and uniquely—by erecting, each according to its kind, a specialized and characteristic structure on the arena. An examination of the nest and eggs of the various bower-birds will not always permit quick recognition of the species which produced them. But a glance at any bower and its decorations will reveal instantly the identity of builder and collector. The bowers are as distinctive as the birds themselves. This specificity, taken in conjunction with the male display and its attraction for the female, undoubtedly heightens reproductive isolation and obviates any chance of hybridization which, as Stresemann[268a] seems to have shown, has given rise to some of the rarer 'species' of birds-of-paradise.

In the evolution of bower-bird behaviour the display came first. The ancestral bower-birds no doubt displayed arboreally as do most perching birds. Next, they probably displayed terrestrially with coloured or otherwise distinctive articles found in the rain-forest. Various other birds—Little Pied Cormorant, Peaceful Dove, Koel Cuckoo, and passerines—will pick up odd things, including nest-material and flower-petals,[46, 110] and use them briefly in display. Although such behaviourisms can scarcely be called inappropriate (in that they probably enhance the display) their genesis is possibly to be found in substitute or displacement[274] activity. However, in such a case as this, when we are dealing with origins—with long-gone events of which we can have no real knowledge—it behoves us to be cautious before committing ourselves unreservedly to any such simple generalization. We can at present suggest merely that if an inherent tendency towards such minor display makes for increased reproductive efficiency by strengthening the pair-bond, or by other means, it should have survival value and become exaggerated.

Once such a habit was established, it is legitimate to visualize ancestral bower-birds selecting various objects during the height of the display-season and posturing with them, in the absence of a bower, in front of the female on the display-ground. Such a relatively elementary performance is carried out by immature male Satin-birds today. When I scattered pieces of blue glass beneath a fig-tree in which young green birds were feeding they descended to the ground, seized the coloured fragments and began an impromptu display. Soon a few inches of earth were bared and several twigs were laid to form a rudimentary arena. The urge to display in the above manner is probably innate. At the moment, however, we can scarcely even guess at how such a mode of behaviour has been incorporated within the central nervous system, ready seasonably to be called into play by the action of sex hormones which are liberated during the maturation of the gonads.

Along with their other peculiar attributes, bower-birds and the Stage-maker have developed also, in varying degrees, a capacity for vocal mimicry to a sometimes astonishing efficiency. Huxley[122] has summarized evidence that in some birds song is innate, whereas in others the young must learn it from older individuals. There is, then, biological value in the capacity of some species to imitate. We know that the practice of vocal mimicry varies remarkably. If this inherent capacity for mimicry, varying between individuals of a species, happens to be of greater value in a particular environment, it is to be expected that natural selection will operate and, as time progresses, a greater proportion of individuals

THE EVOLUTION OF BOWER-BUILDING

with a capacity for mimicry will appear.[167] The selective process should continue for as long as the environmental circumstances in which mimicry has survival value do not change too drastically, or unless other factors alter the habits of the species so as to render mimicry less important, or deleterious, to its breeding or other efficiency.

It had been shown that many rain-forest birds have become 'master mimics' as distinct from 'minor' and 'casual' mimics in the classification of Chisholm.[49] In a thickly wooded environment (in which bower-birds evolved) long-range visual advertisement by means of a distinctive colour pattern cannot be achieved. It is therefore advantageous for an individual to make more and more noise so that rivals for its territory, and members of the opposite sex, will be constantly aware of its whereabouts. Such noise, too, probably helps to strengthen the pair-bond. Some forest-floor species achieve this by the monotonous repetition of a few notes of their own. Birds which inhabit the rain-forest floor (e.g. pittas, log-runners, lyre-birds, megapodes, scrub-wrens, scrub-birds) are notable for the distance at which their calls can be heard. Some of the above, and many others, reinforce their natural notes with sounds borrowed from other birds and objects. This results in the production of a babble of noise that lets every creature in the neighbourhood know the whereabouts of the caller.

It could be claimed that it would be more 'natural' for a species to develop a great variety of calls of its own rather than to use the notes of its neighbours. But it is immaterial what sounds the advertiser uses, so long as it produces a penetrating, fairly continuous babble of noise that makes its situation and disposition known to other members of its species. The Stagemaker, and all bower-birds that have been closely observed, are known to be mimics in varying degree. The Stagemaker sends out an almost continuous stream of mimicry from its singing-stick. Some bower-birds mimic during their display, and most do so in territorial advertisement from eminences near the bower. The two bower-birds whose habits are well known have been found also to employ a stream of excited mimicry as a distraction-display when the nest and young are menaced. Here it would seem that a capacity developed in one set of conditions may have value as a displacement in another quite different situation, but as yet we do not know whether mimicry, in addition to injury-feigning, really succeeds in distracting the attention of predators from the nest.

The Cat-birds

The cat-birds provide an example of how an elaborate terrestrial display (without an actual edifice) can develop in one member of a group while the others retain an arboreal display broadly similar to that of ordinary birds. Thus, one cat-bird, the Stagemaker, has taken partly to the rain-forest floor, cleared a patch of earth, covered it with newly cut leaves placed in a certain way and, from a perch above, sends into the dense mountain forest a stream of melodious noise. In developing this behaviour pattern it has changed also its cryptic, foliage-green, plumage-colour into an equally concealing brown which matches the forest floor, and merges with the shadowy lower reaches of the canopy where it perches on its singing-stick. It has developed also a toothed beak with which it cuts the leaves. In short, it has become *Scenopoeetes*. The display of the bird is no less characteristic than its plumage. It is equally legitimate, if we look at the complete animal rather than a mere cabinet-specimen, to accord it generic rank because of its unique behaviourisms.

The Stagemaker's cleared space is rather like those made by certain birds-of-paradise. But instead of posturing brilliantly with specialized epigamic plumes, it achieves a striking territorial advertisement with a pale pattern of leaves.

We have as yet no real idea of the initial factors which resulted in the remarkable transformation of the habits and physical structure of the Stagemaker. It and the Green Cat-bird occupy the same mountainous rain-forest, breed at the same time of the year, eat much the same things and are therefore almost certainly partial competitors for the food supply. Both the Green Cat-bird and the Stagemaker eat molluscs, but the larger green bird eats small, soft-shelled, arboreal snails whereas the brown Stagemaker devours also the larger, hard-shelled terrestrial varieties which it cracks on a stone on the ground. It is interesting that although *Scenopoeetes* displays on and very near the ground, it builds its nest generally in the topmost limits of the rain-forest, sometimes more than 100 feet aloft. The Green Cat-bird, on the other hand, nests in the sub-canopy, where it gets much of its food.

Ailuroedus, the stock from which *Scenopoeetes* has sprung, occupies a vast range of mountain and lowland country from the Aru Islands, New Guinea, right through the rain-forests of eastern Australia as far as southern New South Wales. *Scenopoeetes*, on the other hand, is confined to a very restricted mountainous area at about the centre of the extended, discontinuous range of the large Green Cat-bird, *A. crassirostris*. It is my belief that the smaller Stagemaker is a development from a stranded remnant of a continental stock of *A. buccoides* affinity which took to the floor of the rain-forest and at the same time evolved its remarkable display. The beaks of *A. crassirostris* and *A. buccoides* each possess far more prominent sub-terminal indentations than do those of bower-birds. Some scores of cat-birds' beaks examined by me each had a prominent notched upper mandible and a slightly less pronounced depression in the lower one where the two structures engage. However distinctive is the Stagemaker's beak (Pl. 25) at first sight, it is in reality not a very profound modification of the structure exhibited in *Ailuroedus* (Fig. 15). The dark colour pattern of *Scenopoeetes* seems readily explicable in the partially terrestrial habits of the bird. We have already, in *Prionodura*, an example of how a continental remnant can survive in a very small, yet suitable, species refuge.

On the other hand, some may think of the Stagemaker as of comparatively recent evolution. It has not been forced to immigrate, nor apparently been accidentally blown by storms (in sufficient numbers to colonize) even to adjacent pockets of rain-forest. Again, unlike *Prionodura*, the Stagemaker has no ground-displaying relatives in New Guinea. The possibility that *Scenopoeetes* is an example of sympatric speciation cannot perhaps be wholly overlooked. It could be suggested that the exploitation of an abundance of snails and insects that live on or near the rain-forest floor may have somehow allowed the development of a new species by means of ecological and behavioural isolation while the parent stock remained in the same locality. However, on the available evidence (as well as theoretically) it would seem that allopatric speciation, possibly from a stranded montane race of the brown-headed *A. buccoides*, is far more likely. The development of sedentary behaviour has been allowed by the never-failing supply of food in the climatically stable upland rain-forest, and thus neither the mountain-loving *Scenopoeetes* nor *Prionodura* has penetrated across the relatively barren intervening lowlands to adjacent 'jungles'.

A classical case of allopatric speciation, or at least incipient speciation, among cat-birds is seen in the separation of *A. c. crassirostris* of southern Queensland and New South Wales from *A. c. melanotus* of Aru, New Guinea, and north Queensland. Here the southern *A. crassirostris* stock has been split by post-Pleistocene continental floral changes and two green forms, slightly distinctive in plumage, voice, eggs, and possibly certain habits, have arisen.

The true bower-birds

We have seen that *Scenopoeetes* illustrates remarkably the development of a display-ground and a display by means other than the bold exhibition of distinctive plumage pattern. The bower-birds and their display have evolved along quite different lines. However diverse the bower-birds appear in regard to plumage pattern, nuptial adornments, and the relative shape, size, and proportion of limbs and beaks, all genera are characterized by the habit of building a bower of sticks on the display-site. Within this habit they have diverged sharply, and quite unmistakably, in at least two directions. One group (*Ptilonorhynchus, Chlamydera, Sericulus*) builds avenue-type bowers. The second group

(*Amblyornis*, *Prionodura*) builds maypole-type bowers. Fragmentary knowledge that has recently come to hand suggests that the monospecific *Archboldia* builds a primitive display-ground unrelated to either of the types above. Thus, a third distinctive type of display-ground probably occurs.

The avenue-builders (Fig. 1) construct, on a basal platform of sticks, two parallel walls of twigs separated by an avenue sufficiently wide for its owner to pass through with comfort. One species adds an additional wall at each end and so achieves three avenues.

The maypole-builders (Fig. 2) all use a sapling as a central supporting pillar. Around this is built a basal cone surmounted by an elaborate structure of sticks. These bowers vary greatly in detail but the maypole feature and the cone are always constant.

The builders of both bower-types use display-objects of various circumscribed colours which are collected with great discrimination, brought to the bower during the period of gonad maturation and occasionally at other times as well. This collection of display-objects, then, is fundamental to both kinds of bower-builders. We have seen that young bower-birds of two species will collect the ornaments characteristic of their kind before they attempt to build bowers. When the Satin Bowerbird changes his bower-site he transports his display-objects along with the first sticks. The basic factor in bower-bird display is sexual exhibition on the specialized territorial region.

After this special form of behaviour was established, the stock branched in the two or three ways outlined above. Members of each group, as we shall see, have radiated farther in individual directions in regard to bower-structure and the selection and collection of display-things.

Although every species, as far as they are known, arranges an absolutely distinctive display-ground it is worth recording that, apart from odd additional walls, several deviations from the characteristic type have been recorded. The Great Grey Bower-bird has built a structure on the roof of a veranda[57] and the Spotted Bower-bird has built an abnormal display-ground on a large broad limb of a tree.[80] North[205] has an account of an even more remarkable bower built by a Spotted Bower-bird. Here a secondary 'bower' is said to have been built above the first (Fig. 17). Chaffer[43a] has information that suggests that an individual male Satinbird built a succession of bowers with additional parallel walls over a period of years. These remarkable variations from normality show how bower-architecture can evolve. We may remember that *C. lauterbachi* has achieved a very different bower essentially by adding two walls to the fundamental plan.

The origins of bower-building are probably to be found in some form

THE EVOLUTION OF BOWER-BUILDING

of displacement activity. Both males and females of most birds possess an inherent urge to build nests. Today only female bower-birds (as far as is known) make nests. These are built of twigs. The males, on the other hand, have taken to building twig structures on the display-ground where they spend much of their time during the sexual season, and which is the focal point of their activity and, along with the female, of their interest. Nest-building among females is controlled essentially by the seasonal liberation of hormones. The same is true of bower-building.

FIG. 17. Abnormal bower of Spotted Bower-bird (*C. maculata*).
(*After Sharpe.*)

Therefore, it would seem that bower-building may have originated as a displacement activity that is fundamentally allied to nest building. In short, nest-building is of a bisexual nature and this has made bower-construction possible in the form of a displaced building drive, the new product of which has become valuable, ritualized, and permanent in the course of the evolution of the species.

At the same time, it is impossible to agree with Soderberg[263] that the bower—his 'play couch'— is an 'imitation of the nest proper' and exercises a breeding stimulus on the female because of the bower's resemblance to the nest. Bowers do not in the least resemble nests. The nests are generally flimsy, saucer-shaped or, in areas of very high rainfall, sometimes rather more substantial, cup-shaped structures built in trees. The bowers, as we have seen, are either decorated pyramids, or towers, or huts, or twin walls built on a flat platform of sticks. Although a few people have sometimes likened the central part of the avenue to a nest, this minor similarity is merely coincidental to the basic architecture.

The round berries or shells, which almost all species choose as part of their decorative scheme, have been likened to eggs. Soderberg tried to make a comparison between the coloured display-objects on the bowers and the 'diversified and richly variegated colours of the eggs'. A glance at bower-birds' display-things and their eggs makes any such hypotheses untenable. Berries and shells are chosen not because they vaguely suggest eggs but because in the rain-forest they are among the most plentiful, distinctively coloured, and shaped objects available. In any case, berries are often absent or, if present, may be swamped by coloured feathers, straw, or flowers, or by bones, and flat or angular pebbles, few of which resemble eggs in the smallest degree.

The modes of action employed by the avenue-builders in bower-building and bower-painting may be mentioned in passing. In species which have been seen building bowers, the neck action is a sliding one of a nature similar to that used by many birds while building their nests even though, of course, the bower-birds are driving sticks into a platform on the ground. While painting, they use a short, jabbing, rather than a brushing, or stroking, movement. I have suggested that bower-painting is displaced courtship feeding. When feeding other individuals many birds use a sharp, jabbing movement. At present, of course, no confident claim can be made that the peculiar sliding motion employed in bower-building, and the jabbing motion used in bower-painting, are significant from an historical point of view.

We have no clue to the reason why one ancestral species took to building two walls on its basal platform of twigs, why another concentrated on weaving a cone of twigs around the base of a sapling, and why a third merely strews a few fern-fronds in a simple arena. We know only that once bower-building got under way it continued in these different directions and that, as species evolved, each created its own modification of a basic type. Each species, too, supplemented the original display-things with others. The males of one species, and perhaps several, came to select display-things coloured like the plumage of the females or their male rivals. (See also Addenda, note 'B', p. 190.)

The maypole-builders

Neither can we say where the various types of bowers first evolved. Avenue-bowers (Fig. 18) are built by all mainland bower-birds except one. The exception is the Queensland Gardener Bower-bird (*Prionodura*) which builds a maypole-bower which, though quite individual, is no more different from the bowers of some members of the New Guinea genus *Amblyornis* than are some of these from one another. Clearly, from a consideration of bower structure, *Amblyornis* and *Prionodura* are of common stock. The continent of Australia and New Guinea have, of course, been united, separated, and reunited on several occasions since the appearance of modern passerine families. At various times there has

THE EVOLUTION OF BOWER-BUILDING

been a continuity of much the same ecological conditions between the two countries. *Prionodura* (now confined to two restricted pockets or refuges of upland rain-forest in north Queensland) indubitably is the isolated remnant of a once more widely distributed maypole-building stock (Fig. 19). In the absence of fossils and more exact knowledge of past changes in climate and floral distribution, we are at present hardly

FIG. 18. Approximate distribution of Avenue-builders.

justified in speculation concerning the precise period at which this particular separation occurred.

Next to the tall central supporting pillar, the most fundamental feature of the maypole bowers is a cone of fibre woven round its base. Throughout rain-forests all over the world occur innumerable slim, straight trees which are branchless and leafless for a considerable distance from the ground. Such characteristic saplings can be seen in Pl. 25 growing in the stage of *Scenopoeetes*. The ancestral maypole-builders probably cleared a space on the rain-forest floor and displayed there with various objects gathered in the forest. Then they took to weaving twigs around the saplings on their display-ground. Fig. 2 shows how *Prionodura* still builds small sub-bowers, in the form of cones, in the vicinity of the principal structure. This isolated Australian member of the group builds a bower that is probably near to the primitive type. Gradually it adds to the original simple basal cone and to a second smaller one that it raises around the base of an adjacent sapling. These two components it links up. Between them is a special display-perch. On the fabric of the bower it plants orchids and mosses. Each decorated unit is still only a

FIG. 19. Approximate distribution of Maypole-builders.
Stripes: Genus *Amblyornis*.
Spots: Genus *Prionodura*.

repeated cone. All other maypole-bowers, whether the decorated tower of *Amblyornis macgregoriae*, or the elaborate garden-fronted huts of *A. inornatus* or *A. subalaris*, are extensions of the basic *Prionodura* pattern. All possess a dwarf basal cone architecturally homologous with the small simple structures that are found near the principal bower of *Prionodura* (Fig. 2).

It is not known if the basal cone concealed beneath the hut of *A. inornatus* is decorated. If not, its retention poses the question of its value. It does not, apparently, support the roof of the hut. It is possible that the central cone in each of the two hut-type maypole bowers has a protective function. If it were not present, a predator, e.g. a marsupial

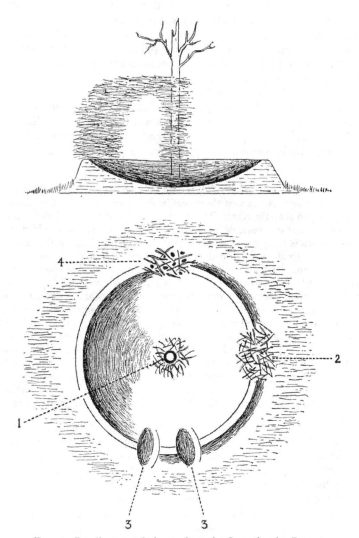

FIG. 20. Peculiar maypole-bower from the Owen Stanley Ranges. (*After Sharpe.*)

1. Central pillar or maypole.
2. Position of arch.
3. Entrance to 'basin'.
4. Position occupied by black decorations.

'cat', native dog, or reptile, could perhaps take stand at the hut-entrance and prevent the bird's escape from within. The presence of the central cylinder compels an attacker eventually to move either right or left within the hut. Thus it would seem to give the confined bird a chance to escape around the other side of the pillar and out into the forest.

There is, in the literature, a nineteenth-century description and sketch (Fig. 20) of a remarkable maypole structure, attributed to *A. subalaris*, that was found in the Owen Stanleys. This bower does not exactly conform to any better-known type. If the sketches are accurate, the bower described may belong to some species which has yet to be described. There is, of course, a possibility that the bower may be aberrant.

The bower in question was first illustrated in figures by De Vis, and later by Sharpe.[256]

This bower [the description runs] is built of twigs arranged in the shape of a hollow circular basin, about three feet in diameter, the side being some six inches higher than the centre. The whole of the basin is covered with a carpet of the greenest and most delicate moss . . . planted there by the bird itself. The surface is scrupulously cleared of all leaves, twigs, &c. In the centre of the basin a small tree, without branches, about two inches in diameter, is growing. Immediately around this tree, and supported by it to about two feet, is erected a light structure of sticks and twigs, placed horizontally and crossing one another. On the extreme outer edge of the basin a more substantial collection of twigs had been built up, which was arched above so as to join the collection around the central pole, leaving a clear space for the bird to pass through in his gambols. The basin has two entrances leading into it. They are four or five inches apart and are formed by a gap or depression in the outer rim. The bower [= raised bridge-like structure] is placed immediately to the right of the entrances. At the opposite side to the entrances, and on the highest part of the raised rim of the basin, is placed a quantity of black sticks (4 inches or so in length), black beans, and the black wing-coverings of large coleoptera. . . .

The above description and figures (p. 175) suggest that the structure, while built on the maypole plan, is clearly different from all other bowers so far described. It contains certain features of the bower of *A. macgregoriae*. Thus, it possesses a rim raised around a circus which surrounds a sapling. However, there is mentioned no decorated basal section of fibre such as occurs beneath the tangled and projecting twigs which interlace the maypole of *A. macgregoriae*. The curious bridge-like section of fabric from the central pole to the raised rim might suggest that the bower belonged to *A. subalaris* and was sketched during the time when the construction of the hut-roof had just begun. But, according to published data, the bower of this bird has a decorated basal cylinder and apparently no pair of special entrances into the circus. Further, the sketch seems to make it certain that the circus is bounded by a parapet rather than a stockade. Again, the situation occupied by the decorations, as well as their uniform black colour, is different from corresponding features in all other published descriptions.

The avenue-builders

The avenue-builders are distributed widely throughout both Australia and New Guinea (Fig. 17). The Satin-bird (*Ptilonorhynchus*) is monospecific and confined to the rain-forests and denser types of open-forest between north Queensland and Bass Strait. The genus *Chlamydera* is composed of four species. Two of these are spread widely over the mainland of Australia. One is restricted to a strip of tropical north-eastern Queensland and the coasts and foothills of New Guinea. One lives in the *kunai* and *pit-pit* grasses of the high interior of New Guinea. The Satin-bird is sexually dimorphic in a most striking way. On the other hand, three species of *Chlamydera* are monomorphic and the fourth (which inhabits the yellow grass-lands of New Guinea) has recently developed a richer harmonizing coloration in the male. The two species restricted to Australia are crested in both sexes, although in one the crest is lacking in perhaps the majority of females.

Despite the above differences, and others that emerge from an examination of skins, the astonishing similarity of the complex display and reproductive cycle of *P. violaceus* and *C. maculata* makes it obvious that the two genera are closely akin and, possibly, of comparatively recent common ancestry. Both build much the same kind of bower, take up territory early and, as the sexual season heightens, refuse to tolerate any other male at the bower. The display to the female is broadly similar in both species. Both paint their bowers. The Spotted Bower-bird has been observed to copulate at the bower and the Satin-bird will copulate with a mounted female beside its bower. (So, too, will *C. cerviniventris*.) The males of both species mimic vocally in the vicinity of the bower, and the males of neither take any part in nest-building, incubation, or the rearing of the young. The nests of both species are very alike and so are the eggs. A post-nuptial flocking phase occurs in both species. The females of both use vocal mimicry as a distraction display. There are numerous other minor similarities. There are, of course, differences. These are connected mostly with the decorations and their placement; further, the strikingly conspicuous male Satin-bird behaves differently, in his heavily timbered environment, from the cryptic-coloured male Spotted Bower-bird which inhabits the sparsely timbered interior.

It is held, then, that *Chlamydera* is a relatively recent off-shoot from *Ptilonorhynchus* stock, and that its radiation into four cryptic-coloured savannah or grass-dwelling species has been consequent upon the desiccation that has swept outwards from central Australia in comparatively recent times. This modern aridity exterminated numerous species, and was responsible for the differentiation of the many xerophilous forms that have inherited inland Australia and many of the drier coastal areas as well. Enough is known about post-Tertiary changes in Australia and its near north to make it clear that during the middle Pleistocene most of the continent was forested and that the top half, the south-eastern coast, and part of the south-western coast of the continent were covered with dense rain-forests.[29, 92] These shrank until, by late Pleistocene, they were confined to a large tract of country extending from the central-eastern Queensland coastline in a southerly, next westerly, and then northerly direction around to central Western Australia. During this period Australia and New Guinea were probably connected for a time by wide grass and savannah tracts. As aridity spread, the rain-forests further diminished. I believe it possible that the ancestors of the present *Chlamydera* stock were isolated in a reduced westerly pocket of rain-forest of much the same nature as those which still support *P. violaceus* on the eastern coast. This pocket gradually disappeared and forced the adaptation of the isolated stock to the dry conditions which were soon to prevail over much of the Australian mainland. Thus *Chlamydera* arose, retaining many of its ancestral behaviourisms, but of necessity changing over to a cryptic sexual monomorphism which enabled both sexes harmoniously to merge with the new, brown, comparatively open-timbered environment which lacks concealing cover and which developed, at the same time, a greater number of predators. Assuming that the ancestor of *Chlamydera* was dimorphic (as are most rain-forest bower-birds), the change to monomorphism is not as radical as might appear at first sight. For example, green 'immature' male Satin-birds reproduce. It seems not unlikely that, in a changing environment, sexual dimorphism might disappear and a new species arise by a process much simpler and quicker than the paedogenesis exhibited in 'lower' vertebrates.

The choice of pale-coloured and reflecting bower-decorations by the Spotted Bower-bird may be governed by the pale-coloured ventral surface and opalescent crest of the female (as the selection of blue and greenish-yellow in the Satin-bird appears to be dictated by the colours of the female).* But in any case, the choice of bleached bones, shells, or otherwise pale display-things is a natural adaptation to the materials available in the new, periodically drought-swept environment.

The *Chlamydera* stock radiated, as has often happened following drastic changes in a landscape. The events that followed are difficult to interpret. On the mainland today we have two endemic species—the Spotted Bower-bird, which ranges widely over the southern half of

* See, however, Addenda, note 'B', p. 190.

the continent, and the larger Great Grey Bower-bird which occupies the northern open 'scrub-lands' which are dry for months on end and are then subjected to torrential monsoonal rains. Both species are crested. These are obviously closely related and recently differentiated: the bigger tropical bird has adapted itself to the hot and humid scrubby forests of the north, while retaining many of the features of display that are fundamental to the *Ptilonorhynchus/Chlamydera* complex. However, before the development of this crested and typically mainland group as we know it today, it seems probable that there diverged a stock which, still crestless, invaded New Guinea. This stock also branched into two species. Thus, today in New Guinea we have the yellow-breasted *C. lauterbachi* which has penetrated into the sunny grass-lands of the interior and which has become dimorphic once more. The male, which displays at its bower amid tall grass, is cryptically coloured with an olive-bronze head, a golden-brown mottled back, and a pale golden breast. If the female is like other females of the genus, she is an unobtrusive watcher beside the bower: certainly she retains a plumage pattern that harmonizes protectively with her surroundings while she is at the nest. (Here we have a very pretty illustration of the axiom that the survival unit of a species is the *pair*.) This species, incidentally, has added another wall at each end of the bower at right angles to the original pair and so it possesses three avenues instead of one. It uses highly coloured berries and pale bluish-grey pebbles as display-objects.

Also in New Guinea is the monomorphic fawn-breasted *C. cerviniventris* which inhabits the open forested coastlines and foothills and retains the simpler single-avenue bower of its mainland relatives. It, too, uses essentially berries instead of shells and bones. It will be seen that neither of these New Guinea birds, though they inhabit a heavily rain-forested country, has yet taken wholly to the rain-forest. *C. cerviniventris* occurs also in the Torres Strait Islands and on a restricted low-lying strip of north-eastern Cape York Peninsula where it is hemmed in on all sides by the Great Grey Bower-bird (*C. nuchalis*). It could be suggested, of course, that *C. cerviniventris* is a recent off-shoot from the *C. nuchalis* that has lost its crest in both sexes and has been forced by the more successful *C. nuchalis* into this small strip of country and across to New Guinea where it has spread successfully and widely. However, in view of its close resemblance to *C. lauterbachi*, its adaptability and success in New Guinea, and its relative abundance on Cape York, it seems probable that it is spreading south, rather than retreating north. Whether the two New Guinea species exhibit a secondary loss of colour in regard to crests, or whether they never developed them, is a debatable point. I do not believe they ever possessed them.

The third genus of avenue-builders, *Sericulus*, is strictly confined to the rain-forest. It consists of the Regent-bird (*S. chrysocephalus*) which extends only from south-eastern Queensland to central-eastern New South Wales, the New Guinea Regent-bird (*S. bakeri*) which has been found as yet only in the coastal mountains of northern New Guinea, and two distinctive races of the beautiful long-plumed Golden-bird (*S. aureus*), also of New Guinea. The two Golden-birds are very different from the Australian Regent, but the New Guinea Regent is clearly intermediate between the Australian Regent and the Golden-birds and makes no doubt whatever of the relationship of all three types. It will be recalled that the head of the Australian Regent-bird is covered with a most curious short, close-packed, and almost moss-like plumage. *S. aureus* of New Guinea, so unlike the Regent in several respects, retains a tiny 'mossy' tract around its eyes: surely a most unusual kind of vestigial structure.

There is no certainty that the bower of any New Guinea *Sericulus* has been discovered, but an unverified report suggests that the bower of *S. bakeri* is very similar to that of *S. chrysocephalus*. It is interesting that the Australian Regent is not confined to the mountain-tops, and neither, apparently, are the New Guinea birds. It would seem that a single species of golden avenue-builder (*S. chrysocephalus*) has been stranded in the remnants of the continental rain-forests.

Comparatively little is known about the bower display and the sexual cycle of the Australian Regent. It is an avenue-builder. Despite its different plumage pattern, its display-habits are basically similar to those of the other avenue-builders of the genera *Ptilonorhynchus* and *Chlamydera*. The male takes up territory early, displays to the female at the bower with objects which seem to resemble certain of her physical aspects. The male paints the walls of his bower. The nests and eggs are remarkably similar. Only the female builds the nest, incubates, and looks after the young. There is a post-nuptial flocking phase and a plurality of birds is permitted at the bower early in the season. Because of these undoubted similarities, the view that *Sericulus* is 'apparently of no relationship with the other birds with which it is associated merely on account of its bower-building habits'[185] is rejected.

We cannot, from the facts at our disposal, say that the avenue- or the maypole-builders originated in one country or the other. But it is possibly significant that the only maypole-builder on the mainland is *Prionodura* which occupies a couple of restricted refuges in tropical Queensland in an area which also supports a solitary patch of rhododendron which is also unknown elsewhere on the Australian mainland. According to David,[65] the rhododendron probably migrated south to Australia during a Pleistocene period when the two countries were united. The same may be true of *Prionodura*. At the moment, however, it only seems certain that the family as a whole arose in the thickly forested Australo-New Guinea area after its separation from the Malayan region. To surmise further would involve arguments that are far too hypothetical to be legitimate in the present study.

The classification of bower-birds

The habits of the birds themselves have never been taken into account in any of the several attempted classifications of the Ptilonorhynchidae even though their enormously exaggerated behaviourisms make them ideal subjects for such study.* In fact, it is probable that it is the very development of these complex behaviour patterns that has been partly responsible for the physical differences among bower-birds that have been so baffling to certain cabinet-workers in the past. The establishment of exaggerated patterns of display probably leads to different evolutionary rates and causes even wider variation of the physical structures connected with display, and possibly other characters as well.

There is ample evidence that bower-form is a more constant and reliable diagnostic feature than certain prominent physical structures. Take, for example, the genus *Amblyornis*. *A. inornatus* is drab in both sexes, lacking the bold, flaring crest that typifies the male of each other member of the genus. The crestless *A. inornatus* and one crested bird (*A. subalaris*) both build huts. Another crested bird (*A. macgregoriae*) merely decorates a central tower. But all three weave a basal cone around a central maypole. Here we have an interesting example of a behaviourism remaining relatively constant while a most striking morphological character does not. The same is true also among avenue-builders. We saw in the genus *Chlamydera* how there has probably been a loss of a striking sexual dimorphism and the development of a comparatively drab monomorphism, with the addition of a bright, erectile, and epigamic crest, in a relatively short space of time. In this genus we saw also the evolution of a secondary sexual dimorphism, involving a gain of colour in the male, when a stock (*C. lauterbachi*) again changed its environment. And, from our study of the behaviour of the *Ptilonorhynchus/Chlamydera* group, we have found how close in reality the members of two seemingly widely different (but quite artificially erected) subfamilies can be.

These examples provide a cogent warning against placing too much reliance on epigamic and other superficial physical features (the very nature of which may make for their comparatively rapid development) as a guide to supra-specific, and particularly supra-generic, relationships.

* See, however, Addenda, note 'F', p. 192.

It is of high interest that *A. inornatus*, the only maypole-builder in which the male is drab and coloured like the female, makes what is reputed to be the most ornate and structurally elaborate display-ground of all. Neither it nor two of the four species of *Chlamydera*—*C. cerviniventris* and *C. lauterbachi*—has yet developed a crest. Unless each of these species has entirely lost such striking epigamic features—and this is surely unlikely in the males—it would appear that bowers, and behaviour patterns, of great complexity developed first, and that the crests came later as a response to the selection pressure engendered by the display at them.

From the study of gross external anatomy, bower-birds have been said to show 'little inter-related connections'; and it has been suggested also that 'bower-making has been developed coincidentally or perhaps independently'.[126] Another modern writer[185] came to the conclusion that 'the bower-building habits seem to be imitative and adaptive, and ... do not indicate close relationship, four very distinct groups being easily recognisable, with probably three sources without affinity'.

I have made it clear that such views cannot be sustained if we consider bower-birds as living animals outside the museum cabinet. Further, if bower-building arose independently, it might not be too much to expect that the habit could, though of course not necessarily should, have arisen in similar ecological conditions in not entirely unrelated birds living in South America, Africa, and southern Asia. After all, birds as unrelated as grouse and birds-of-paradise build simple display-grounds; and various anatomical and physiological characters (for example, vivipary) have arisen independently in animals not only of different families, but even classes.

It has been said, too, that certain bower-birds would never have been associated in classification 'save for the bower-building habit, and this is not a phylogenetic character, so later students will also dissipate the members of this so-called family elsewhere'.[126] From the present study it is held that, contrary to the above opinion, bower-building is of the greatest phylogenetic significance. Further, if the various avenue-builders (for example) would not have been placed near each other save for the bower-building habit, it leads us to wonder just how much worth can be attached to some classifications in the realms of supra-generic systematics. The latest such classification has been made by Iredale[126] (p. 231) who divides the Ptilonorhynchidae into the following sub-families and genera:

Subfamily Sericulinae: Anomalous bower-birds
  Genus *Sericulus*
  Genus *Xanthomelus*
  Genus *Cnemophilus*

THE EVOLUTION OF BOWER-BUILDING

Subfamily Amblyornithinae: Gardener Bower-birds
  Genus *Amblyornis*
  Genus *Prionodura*
  Genus *Archboldia*
Subfamily Ailuroedinae: Cat-birds
  Genus *Ailuroedus*
  Genus *Scenopoeetes*
Subfamily Ptilonorhynchinae: Satin Bower-birds
  Genus *Ptilonorhynchus*
Subfamily Chlamyderinae: True Bower-birds
  Genus *Chlamydera*.

The above arrangement is untenable in the light of the data presented in the foregoing chapters. For example, the wide separation of Iredale's so-called 'true' bower-birds (*Chlamydera*) from his 'anomalous' bower-birds (*Sericulus*) (both avenue-builders with a broadly similar behavioural cycle) is completely artificial in that both, according to their behaviourisms, developed from a common stock and most assuredly should not have a group of maypole-builders and the cat-birds (which do not build bowers at all) interposed between them. Again, if on purely structural grounds the 'anomalous' bower-birds are given subfamily rank, it would seem that the same at least should have been accorded *Cnemophilus* which is almost certainly not a bower-bird at all. And although the cat-birds and the avenue-builders may have arisen from a common stock at some remote period, it is difficult to follow the reasoning that the Satin Bower-bird in particular is an 'abnormal off-shoot of the cat-birds'.

With the exception of *Archboldia*, bower-birds and cat-birds appear to fall naturally into three groups. These are as follows:

1. Maypole bower-builders (*Prionodura*, *Amblyornis*).
2. Avenue bower-builders (*Sericulus*, *Ptilonorhynchus*, *Chlamydera*).
3. Cat-birds (*Ailuroedus*, composed of almost exclusively arboreal birds which build no display-grounds as far as is known, and *Scenopoeetes*, an off-shoot which has taken partly to the rain-forest floor, and builds an extensive display-ground which it covers daily with fresh leaves during the sexual season).

It is essential to adopt a conservative attitude in regard to *Archboldia*. So little is known of its behaviour. Likewise, we know precisely nothing about the habits of *Amblyornis flavifrons*, another bird with a curious though different crest. At present I incline to the belief that *Archboldia*, instead of being closely related to *Amblyornis*, as has been suggested, represents a distinct stem of its own. It is difficult to believe that its peculiarly simple display-ground represents the degeneration of a more complex structure of the maypole type. It might be suggested that *Archboldia*, by virtue of its frond-strewn arena, could be closely allied

184 THE EVOLUTION OF BOWER-BUILDING

to *Scenopoeetes*. But I see no reason for adopting such a view since their display-grounds appear to have little in common. The affinities of these birds will be settled only by the employment of more refined tools (probably biochemical ones) than are at our disposal at the present time.

There seems to be no good reason for including the cat-birds in the

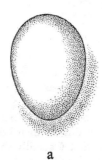 

a   b

FIG. 21. Eggs of bower-birds.
Left: a. Unmarked eggs of the maypole-builders (so far discovered).
Right: b. General pattern of the eggs of the avenue-builders. (Those of *P. violaceus* are usually more spotted than streaked.)
(The eggs of *Archboldia* are still unknown.)

same family as the bower-birds. The most logical arrangement for the entire group may be as follows:

Family Ptilonorhynchidae (bower-birds)
   Sub-family Amblyornithinae (maypole-builders)
   ? *Archboldia*
   Sub-family Ptilonorhynchinae (avenue-builders)
Family Ailuroedidae (cat-birds)

It is believed that the above simple classification will be shown, in the light of future research, to be a truer one than those based on the mere consideration of external anatomy and colour pattern which, however reliable for the determination of specific and generic rank, is apt to be grossly misleading in appreciations of remoter relationships.

It is of outstanding interest that, even if the maypole-builders and the avenue-builders do not exhibit physical characters that render them readily divisible, their eggs, as far as they are known, do so to a most striking degree (Fig. 21).* All avenue-builders whose nests have been discovered (six of a total of eight species) produce eggs that are streaked and zigzagged and sometimes spotted in a most peculiar and quite characteristic manner. So far, the eggs of perhaps only three maypole-builders are known. These are all unadorned white or yellowish-white.

* See also Schönwetter[249a] for a detailed account of the eggs of both bower-birds and birds-of-paradise.

The eggs of *Scenopoeetes* and the two cat-birds are also unstreaked, but it is not at present suggested that this necessarily indicates close relationship with the maypole-builders. All kinds of species of the most diverse relationship lay white or whitish eggs.* The eggs of *Archboldia* are still undescribed.

Intelligence and aestheticism in bower-birds

Many people have studied bower-birds and failed to appreciate the utilitarian significance of bower-building, colour-display, and painting, and therefore assumed that the birds must intelligently carry out these activities as a kind of relaxative hobby and to express a unique sense of artistry and aestheticism. To some degree the above notion is an exaggerated twentieth-century echo of the view put forward by Romanes[241a] not long after the death of Darwin. Romanes, while admitting the 'incalculably great . . . explanatory value of the Darwinian Theory of Natural Selection', believed that it could not account for 'all that class of phenomena which go to constitute the Beautiful' since, he said, 'whatever value beauty, as such, may have, it clearly has not a life-preserving value'. Romanes cited particularly the bower-birds—'the most consummate artists'—in which it is 'impossible to doubt that an aesthetic sense is displayed' and in which, he asserted, bower-building and associated phenomena 'have no reference to utility or the preservation of life'.

While I have ascribed a utilitarian basis for each of the behavioural phenomena discussed, I see no reason, provisionally, to deny that bower-birds possess an aesthetic sense although, it must be emphasized, we have as yet no concrete proof that such is the case. Some bower-birds certainly select for their displays objects that are beautiful to *us*. Further, they discard flowers when they fade, fruit when it decays, and feathers when they become bedraggled and discoloured. But, it must be remembered, however beautiful such articles may be, they are still probably selected compulsively in obedience to the birds' heredity and physiology. Even though some species exhibit slightly varying preferences in shades and shapes, that which habitually chooses an essentially deep blue and lemon-yellow decorative scheme, for example, does not occasionally vary its decoration to red and white, like a woman arranging flowers, nor even to pale blue and gold. The choices, in the species we know best, are mechanical; and so, seemingly, are the other bizarre activities which have excited so much imaginative writing in the past. The fact that some bower-birds select objects that appeal to man's sense of beauty is no proof that such articles have a similar effect on the bird. If all bower-birds made collections of bleached bones, less would be written of

* See, however, Stresemann, Addenda, note 'F', p. 192.

aestheticism. Yet nobody would suggest that its pile of dry bones and dead snail-shells is less beautiful to *Chlamydera nuchalis* than is the 'beautiful' array of blue and red berries to *C. lauterbachi*. It would, of course, be unthinkable to suggest that bower-birds—or any birds for that matter—do not get pleasure from the vocal, architectural, and other activities they perform, but whether such pleasure has much in common with that of Man, engaged in comparable pursuits, has yet to be proved.

The question of intelligence is perhaps a little less complicated one. Certainly it is susceptible to experimentation. We have seen that the Satin-bird can learn quickly from experience and remember a brief lesson, concerned with its welfare, over the space of a year. But it has to be emphasized that there is no experimental evidence whatever that bower-birds are any more intelligent than other highly organized passerines. Quite apart from their remarkable display-habits *per se*, bower-birds—and merely because they are bower-birds—have been often credited with intelligence when carrying out activities common to other quite unremarkable species. Thus, 'bower-birds are clever from the day of their birth', asserted one observer[91] merely because the young of the spotted species clung tenaciously to the nest with their claws and later shammed death; and because, too, the female practised distraction-display. Another writer[134] ascribed 'foresight' to the same species because it builds its bower under the shelter of low-hanging branches where it is protected from the trampling hooves of cattle. The desire to build among the shadows of low-sweeping boughs is traditional and innate, and no more an expression of reasoned behaviour than is the urge of a Robin to conceal its nest in an ivy-covered chink in a wall. Bower-birds build within the shady protective concealment of thickets in areas that have never been inhabited by livestock.

I have described in previous chapters bower-bird behaviour which, judged by the standards of a sapient animal, could be considered extraordinarily stupid. An example is the inability of the Stagemaker to realize that a sharp downward sweep of its neck would, with some leaves, achieve in a second a result that it takes minutes to bring about with the laborious sawing motion of its beak. It has the nervous and muscular equipment with which to perform this simple act, but it persistently carries on the complicated, arduous, and time-consuming sawing activity that has evolved along with its specialized 'toothed' beak.

However, it is as unfair, by our standards of reasoning, to call the bird stupid, as it would be for it, if it were able, to call *Homo* stupid because he lacks the Stagemaker's agility in the tree-tops. It is scarcely necessary to say that both Man and bird have progressed a long way along widely divergent paths in their individual evolution.

We can possibly learn something concerning the origins of certain behaviour of Man from a study of the innate, unconditioned reactions of bower-birds, but it is dangerous to project anthropomorphically upon bower-birds certain attributes that have come to Man as a result of the hyper-development of his cerebral cortex. Although various writers have uncritically tried to invest bower-birds with human attributes, it is equally appropriate to speak of bower-bird-like habits within the Hominidae. After all, the Ptilonorhynchidae probably evolved, and were displaying, before *Homo sapiens* appeared on earth. The desire to display and to exhibit associated aspects of the sex drive is common to both the Aves and the Mammalia (including Man) and is allowed essentially by the action of sex hormones on specialized centres within the central nervous system. The clinical literature abounds in cases in which, for example, either lesions of the hypothalamus, Leydig cell tumours, or neoplasms of the adrenal cortex have resulted in the production of abnormal amounts of male hormones which have in turn caused precocious puberty and embarrassingly overt sexual behaviour in little boys only a few years old. Although in modern Man sexual behaviour is enormously complicated by psychogenic influences, it is probable that dancing, with its rhythm and accompanying music-making, originated in our remote ancestors as a form of sexual display. The violence that is still associated with the dances of many so-called primitive peoples suggests that in its even deeper origins it may have a basis, in terms of ethology, in displacement activity. Display in both birds and Man probably possesses an aphrodisiacal quality. Further, some human gestures and attitudes in both male and female have, in an almost equally simple, conspicuous, and specific way, much the same functions as have the sign stimuli which for generations have been observed, and are now being understood, in fishes, amphibians, reptiles, birds, and the sub-human Mammalia.[6, 153, 274] In most vertebrates, however, it would seem that such behaviour patterns are largely or wholly inherent and, as we have seen in bower-birds, controlled in their expression almost entirely by internal and environmental stimuli, including those from the opposite sex and from other members of the community. Their operation can be inhibited by internal factors (including sickness) as well as by external stimuli such as cold, hunger, predators, and the seasonally recurring absence of conditions to which the creatures traditionally respond. Likewise, in Man, the urge to indulge in sexual display and concomitant activities is probably still to some degree innate and is certainly similarly influenced by internal, and environmental, stimuli and inhibitors. But, in powerful addition, Man's supremely developed cerebral cortex has led to his endowment with critical and moral faculties. These include, in each 'normal' person, the capacity to see beyond the here and now, the ability to control instinctive drives and to redirect them into activities that are considered desirable, or at least acceptable, by the particular community in which he dwells.

23

ADDENDA

p. 35. 'A': *Temperature and bower-building*

At Bart's in 1953 the first autumn bower was built (by a pair of non-experimental birds kept together with a view to breeding) on 4 September (= 4 March in Australia). It was demolished a few days later. The birds displayed spasmodically on perches until 10 December (= 10 June), when a new bower was made. This was used until 20 December, when, in a period of 4 days, the minimum temperature dropped from 47° to 35° F. and the bower was demolished. There now came immediately a sharp rise in temperature (35° to 48° in 24 hours) and the wrecked bower was rebuilt. A second pair kept at the London Zoo for breeding purposes built a new bower on the same day and displayed vigorously. Both bowers appeared to be permanent structures (although it is impossible to be sure of this under captive conditions) and were constantly used until the first week in January when temperatures dropped to 33° and the Bart's bower was pulled to pieces. The Zoo bower remained intact, but was not used. This bower was refurbished on a particularly warm day (maximum 58°, minimum 48°) in mid-January, but was again neglected during three weeks of severe weather when minimum temperatures ranged from 22° to 30° and maximum temperatures varied from 27° to 41°. During this period arboreal display, too, almost ceased among both pairs. On 9–10 February the maximum temperature rose from 42° to 54°, but the minimum was only 35°. The Bart's bird still did not rebuild. Two days later, when the maximum reached 53° and the minimum rose to 45°, the bower was rebuilt and was continually used in the spring-display of 1954.

The above data show that there is a tendency for bower-building to be renewed at any time during the winter (including before the solstice) as soon as temperatures rise to a level near that to which the birds are adjusted during the winter at home (pp. 33–34). This of course does not mean that the seasonal gametogenesis is under temperature control. It is possible, however, that prolonged sunny weather and consistently high temperatures will stimulate bower-building and display and, after the refractory period, may thus indirectly initiate seasonal sexual changes and the breeding season irrespective of day-length. But it must be remembered that none of the above birds was examined histologically, and so it was not known whether seasonal spermatogenesis had begun. It is possible that both the captive birds whose display was influenced by weather fluctuations may have been at all times still in the pre-gametogenetic phase and that some additional factor such as increasing daylength was necessary to start their seasonal spermatogenesis. In short, while there is no adequate proof that light *per se* stimulates the breeding rhythm outside the laboratory, there is certainly no positive proof that it does not. The sacrifice of 15 displaying birds between 15 and 21 June in their native forests during an unusually warm and sunny winter and a similar number in a normal year would do much to settle this perplexing point. Dozens are wantonly slaughtered a little later each year by orchardists and others.

pp. 49, 64, 118. 'B': *Factors governing colour selection*

Experiments made in the spring of 1954 suggest strongly that the selection of the peculiar colour of greenish-yellow (pp. 37, 48), and perhaps the blue, is dictated by the coloration of rival *males* and not that of females as has been previously thought. The death of an old blue male made it possible to experiment with its fresh, unfaded greenish-yellow-tipped beak. Blue males instantly collected the disarticulated mandibles, carried them to the bower and displayed vigorously to females. They ignored the dark, horn-coloured beak taken from a female. The small greenish-yellow mandibles of the blue male were used as readily as blue test objects twice their size. Both greenish-yellow beak and blue test objects were preferred to yellow feathers taken from the skins of females and younger green males, which, however, were also taken to the bower for display. Black, 'blue'-edged, yellow-vaned feathers from the skins of blue males were taken to the bower much less readily than the former articles, and less readily still if the pale yellow basal section of the vane was first removed. After such mutilation, feathers from blue males were often neglected for hours, but were eventually brought to the bower. Pale yellow, brown-barred feathers (from the under wing-coverts) of females were neglected altogether.

In a brief war-time test the somewhat faded beak taken from a formalin-preserved blue male was ignored by a displaying blue bird. After the experiments recorded above, it was thought worth while to duplicate the war-time test by presenting two faded (but still distinctly greenish-yellow) preserved beaks to the experimental birds that had accepted the detached fresh beak for display. One very faded beak was ignored. The second slightly less faded specimen was taken to the bower only after four days.

The greenish-yellow of the beak (about 'light greenish-yellow' of Ridgway's *Color Standards*[231]) is comparatively rare in nature and is found in few local wild flowers or other objects. It occurs only very approximately and inconspicuously in the female plumage. Although the above experiments must be repeated and elaborated, it seems that we now see the underlying basis of the Satin-bird's curious urge to collect the small, pale and relatively inconspicuous bells of *Billardiera* and its preoccupation with pieces of onion scale-leaves only after their decomposition to a similar shade of greenish-yellow (p. 38).

It is possible that it is the blue eyes (p. 63) and the small cone-shaped area (6 mm. long) of greenish-yellow exhibited by the exterior and interior of the beak-ends (and the thin beak-edging of the same colour) that are the principal sign-stimuli significant in short-range relationships between rival males. That may be why blue and greenish-yellow are the predominant colours taken to the display-ground and used there in the aggressive display which, as we have seen (pp. 62–64), is partly a device to keep rival males out and the attached female in.

It now seems not unlikely that in addition, the shades of 'pure' yellow, brown, and grey (in straw, snail-shells, puff-balls, snake-skins, &c.) are chosen not because they match the female's plumage colours, but instead because they match those of the nearly similar younger males. Caution, however, is needed here. Although young green males take over bowers and females on the death of the blue owners (p. 45), they are always submissive to blue birds when they meet and do not constitute the marauding threat of rival blue males which constantly invade each others' territories (and of course those of green males), to destroy and steal (p. 46). However, green males possibly steal each others' display-things and perhaps defend territories and so compete among themselves for females. This sub-adult competition may explain the selection of the yellow, brown, and grey by males of all ages.

Although it now seems probable that some coloured objects (greenish-yellow and blue) chosen by the male are selected in the image of rival blue males, and that the remainder are possibly chosen to match the plumage of younger green males, this does not necessarily wholly invalidate the original hypothesis.[199, 209] The colours are still like those of the female, and whatever the factors underlying their original selection, it is not disproved that they may exercise an exciting effect on the male by virtue of the female resemblance.

However, it now seems likely that, in their similarity to prominent physical characters of other males, the decorations provide a vehicle, so to speak, for displaced combat activities. The conclusions[162, 178] concerning the functions of the noisy display with the colours in connexion with territory domination, pair establishment, bower protection, sex synchronization, and energy absorption remain unmodified.

It would seem that the same factors may underlie the conservative selection of coloured display-things by the two endemic Australian members of *Chlamydera* (pp. 87 and 99) and *S. chrysocephalus* (p. 118).

p. 65. 'C': *Antiquity of bower-painting*

Stresemann[268b] believes that the painting habit 'is a late acquisition of *Ptilonorhynchus*'. He suggests that it is found only in the large southern sub-species, since no painted bower has been recorded from the small sub-species in north Queensland. It will be important, in the future, to learn if the northern *P. minor* paints. So far less than half a dozen bowers have been examined to this end (see pp. 65, 117).

p. 70. 'D': *Effect of age or infirmity on display*

A blue male Satin-bird that was trapped in the act of tearing down another's bower in 1949 lost an eye in conflict while housed at the Zoo shortly after arrival in London. It remained apparently healthy, but made no attempt to build a bower although other males in identical cages did so regularly. A secondary indication of age (in addition to its blue plumage at the time of capture) was an overgrown beak from which $\frac{1}{4}$ inch was clipped when the bird was removed to Bart's in May 1953. There, in undisturbed conditions, it still made no attempt to build a bower. In March 1954 it escaped and had to be shot. Seminiferous tubules were clear of lipids and contained mitotic figures and spermatocytes. The interstitium was heavily lipoidal. There were indications of 17-ketosteroids in the testis.[238a] Thus, there was plenty of evidence that the reproductive apparatus was functional: it seemed that infirmity, old age, or undetermined psychical factors were responsible for its failure to build and display (see pp. 45, 70).

p. 151. 'E': *Differences between cat-birds and bower-builders*

Salmon[246a] has described distraction-display by the Green Cat-bird near its nest and young. Unlike bower-building birds, *both* parents performed it. Neither indulged in vocal mimicry as do females of certain bower-birds in similar circumstances (see p. 151).

pp. 181, 185. 'F': *Evolution of bower-building*

Stresemann[268b] believes bower-birds and cat-birds may be divided into two groups according to the colour of their eggs. These are:

1. The Ailuroedinae, containing the genera *Ailuroedus, Scenopoeetes, Archboldia, Amblyornis* and *Prionodura*;
2. The Ptilonorhynchinae, containing the genera *Sericulus, Xanthomelus, Ptilonorhynchus* and *Chlamydera*.

ADDENDA

Stresemann further says:

The hypothetical foundation of the pairing ground of all bower-birds is seen to be a carefully cleared area. The next stage is: the area is decorated, in the case of *Scenopoeetes* with fresh leaves from a tree, and in the case of *Archboldia* with dried fern fronds. Additional structures serve primarily to make the place more striking. They are made by horizontal stacking of twigs in the case of *Amblyornis* and *Prionodura*, and by ramming in vertically supported twigs in the case of *Sericulus, Xanthomelus, Ptilonorhynchus*, and *Chlamydera*. Both methods are apparently an old inheritance. (See p. 181.)

pp. 8, 9, 11, 12. 'G': *Internal events in avian reproduction*

Experimental work by Benoit and associates in France, and by Greeley and Meyer in the United States, suggests that, in the domestic drake and Pheasant (*Phasianus*) respectively, a seasonal period of refractoriness intervenes in the pars distalis as well as in the testes. For information on this, and references to other contemporary work on avian reproduction, see reviews by Marshall (on the male), Breneman (on the female), and reports on experimental studies by Witschi (vertebrate gonadotrophins), Fraps (varying effects of sex hormones), and Benoit (possible humoural activation of the adenohypophysis) in *The Comparative Physiology of Reproduction* (1955), London, a Memoir of the Society for Endocrinology edited by I. Chester Jones and P. Eckstein.

REFERENCES

1. ALLARD, H. A. (1940). *Journ. Wash. Acad. Sci.* **30**, 34.
2. ALVERDES, F. (1932). *The Psychology of Animals.* London.
3. ANON. (1940). *Sydney Morning Herald*, 30 January.
4. ANON. (1952). *Brit. Med. Journ.* **2**, 197.
5. ARCHER, J. A. (1926). *Emu*, **26**, 138.
6. ARMSTRONG, E. A. (1947). *Bird Display and Behaviour.* London.
7. ASSENMACHER, I., GROS, C., BENOIT, J., and WALTER, F. X. (1950). *C.R. Soc. de Biol.* **144**, 1107.
8. *Australian Encyclopaedia* (1925). Sydney.
9. BAILEY, H. F. (1934). *Emu*, **33**, 186.
10. BAKER, J. R. (1938). *Evolution: Aspects of Evolutionary Biology.* Edited by G. R. DE BEER. Oxford.
11. —— (1947). *Journ. Linn. Soc. Lond.* **41**, 248.
12. —— and BAKER, I. (1936). Ibid. **29**, 507.
13. —— and BIRD, T. F. (1936). Ibid. **40**, 143.
14. —— MARSHALL, A. J., and HARRISSON, T. H. (1940). Ibid. **41**, 50.
15. —— and RANSOM, R. N. (1938). *Proc. Zool. Soc. Lond.* A, **108**, 101.
16. BARNARD, H. G. (1911). *Emu*, **11**, 17.
17. BECCARI, C. (1878). *The Gardeners' Chronicle*, 16 March, 332; also (1855), Ibid. 9 May, 695. (Reprint of extract, with illustration.)
18. BERNEY, F. L. (1927). *Mem. Q'ld. Mus.* **9**, 194.
19. BARRATT, C. (1915). *Emu*, **15**, 176.
20. BARRINGTON, D. (1773). *Phil. Trans. Roy. Soc. Lond.* B, **63**, 442.
21. BENOIT, J., ASSENMACHER, I., and WALTER, F. X. (1950). *C. R. Soc. de Biol.* **144**, 573.
22. BERTLING, A. E. L. (1904). *Avic. Mag.*, N.S. **2**, 235.
23. BEST, C. H., and TAYLOR, N. B. (1945). *The Physiological Basis of Medical Practice*, 4th ed. London.
24. BISSONNETTE, T. H., and WADLUND, A. P. R. (1932). *Journ. Exp. Biol.* **9**, 339.
25. —— (1937). *Wilson Bull.* **49**, 241.
26. BURROWS, H. (1949). *Biological Actions of Sex Hormones*, 2nd ed. Cambridge.
27. BONAPARTE, L. (1854). *Ann. d. Sc. Nat.*, Ser. IV., Zool. 122.
28. BOURKE, P. A., and AUSTIN, A. F. (1947). *Emu*, **47**, 110.
29. BROWNE, W. R. (1945). *Proc. Linn. Soc. N.S.W.* **70**, v.
30. BROADBENT, K. Cited by CAMPBELL, A. J., 36.
31. BULLOUGH, W. S. (1942). *Phil. Trans. Roy. Soc.* B, **231**, 165.
32. BURGER, J. W. (1949). *Wilson Bull.* **61**, 211.
33. BURKITT, J. P. (1935). *Brit. Birds*, **28**, 322.
34. BLANCHARD, B. D. (1941). *Univ. Calif. Pubns. in Zool.*, **46**, 1.
35. CAMPBELL, A. G. Cited by MATHEWS, G. M., 185.
36. CAMPBELL, A. J. (1901). *Nests and Eggs of Australian Birds.* Sheffield.
37. —— and BARNARD, H. G. (1917). *Emu*, **17**, 136.
38. CARNABY, I. C. (1946). *W. Austr. Bird Notes*, **4**, 11.
39. CHAFFER, N. (1926). *Emu*, **26**, 24.
40. —— (1931). Ibid. **30**, 277.

REFERENCES

41. CHAFFER, N. (1932). *Emu*, **32**, 8.
42. —— (1945). Ibid. **44**, 162.
43. —— (1949). Ibid. **49**, 19.
43a. —— Personal communication.
44. CHAPIN, J. P. (1929). *Am. Mus. Nov.* **367**, 1.
45. —— (1950). *Abstr. Papers 68th State Meeting A.O.U.* 8.
46. CHISHOLM, A. H. (1929). *Birds and Green Places.* London.
47. —— (1936). *Emu*, **35**, 211.
48. —— (1945). Ibid. **44**, 183.
49. —— (1946). *Nature's Linguists: A study of the riddle of vocal mimicry.* Melbourne; also (1937), *Ibis*, Oct. 703.
50. —— (1949). *Emu*, **49**, 60.
51. COLE, C. F. (1910). Ibid. **9**, 236.
52. COLEMAN, W. H. Personal communication.
53. CONDON, H. T. (1946). *Emu*, **45**, 210.
54. COOMBS, C. J. F. Personal communication.
55. CORNWALL, E. M. (1903). *Emu*, **2**, 233.
56. —— (1904). Ibid. **3**, 188.
57. —— Cited by CAMPBELL, A. J., 36.
58. COUES, E. (1891). *Auk*, **8**, 115.
59. CRANDALL, L. S. (1931). *Paradise Quest.* New York.
60. —— (1932). *Bull. N.Y. Zool. Soc.* **35**, 65.
61. D'ALBERTI, M. L. (1877). *Ibis*, 4th ser., **1**, 363; also *Sydney Mail*, 24 Feb. 1877.
62. DAMSTE, P. H. (1947). *Journ. Exp. Biol.* **24**, 20.
63. DARLING, F. C. Cited by SHANKS, D. (1949), *Emu*, **49**, 132.
64. DARWIN, C. (1871). *The Descent of Man and Selection in Relation to Sex.* London.
65. DAVID, T. W. E. (1932). *Explanatory Notes to Accompany a New Geological Map of the Commonwealth of Australia.* Sydney.
66. DAY, W. S. Cited by NORTH, A. J., 205.
67. DE VIS, C. W. (1883). *Proc. Linn. Soc. N.S.W.* **7**, 562.
68. —— (1890). *An. Rep. Brit. N.G.*, 1888–9, 61.
69. —— (1890). *Proc. Roy. Soc. Q'ld.* **6**, 247.
70. —— (1892). *Ann. Q'ld. Mus.* **2**, 9.
71. —— Cited by SHARPE, R. B., 256.
72. DIGGLES, S. (1867). *The Ornithology of Australia.* Brisbane. Also GOULD, J. (1869), *The Birds of Australia*, Suppl. I.
73. D'OMBRAIN, E. A. (1921). *Emu*, **21**, 59.
74. DUNCAN, M. Personal communication.
75. EDWARDS, H. V. (1920). *Emu*, **19**, 306.
76. ELLIOT, A. J. Ibid. **38**, 30.
76a. ELTON, C. *Voles, Mice, and Lemmings: Problems in population dynamics.* Oxford.
77. EMLEN, J. T., and LORENZ, F. W. (1942). *Auk*, **59**, 369.
78. ERHARDT, H. (1934). *Zool. Jahrb. Abt. Allgemein.*, Zool. **4**, 489.
79. EVANS, A. H. (1899). *Birds. The Cambridge Natural History.* Cambridge.
80. FAVALORO, N. (1940). *Emu*, **39**, 273.
81. FIELD, B. Cited by NEWTON, A., 204.
82. FINSCH, O., and MAYER, A. B. (1886). *Ibis*, 5th ser. **4**, 237.
83. FLEAY, D. (1943). *Avic. Mag.*, 5th ser. **8**, 123.

84. FRANCIS, W. D. Personal communication.
85. GANNON, G. R. (1930). *Emu*, **30**, 39.
86. —— (1936). Ibid. **36**, 361.
87. GASTEEN, W. Personal communication.
88. GAUKRODGER, D. W. Personal communication to CHISHOLM, A. H.
89. —— Cited by MATHEWS, G. M., 185.
90. —— (1922). *Q'ld. Nat.*, April, 81.
91. —— Cited by CHISHOLM, A. H., 46.
92. GENTILLI, J. (1949). *Emu*, **49**, 85.
93. GILBERT, P. A. (1910). Ibid. **10**, 44.
94. —— (1928). *Aust. Zool.* **5**, 141.
95. —— (1939). *Emu*, **39**, 18.
96. —— (1940). Ibid. 209.
97. GILLIARD, E. T. (1953). *National Geogr. Mag.* **103**, 421.
98. GODDARD, M. T. (1947). *Emu*, **47**, 73.
99. GOODWIN, A. P. (1890). *Ibis*, 6th ser. **2**, 150.
100. GOULD, J. (1836). *Proc. Zool. Soc. Lond.* 106 and *Syn. Bds. Austr.* (1837), 1.
101. —— (1840). *Proc. Zool. Soc. Lond.* 94.
102. —— (1848). *The Birds of Australia*, **4**. London.
103. —— (1850). *Proc. Zool. Soc. Lond.* 201.
104. —— (1865). *Handbook of the Birds of Australia.* London.
105. GREEN, B. (1910). *Emu*, **9**, 247.
106. GWYNNE, A. J. (1937). Ibid. **37**, 76.
107. HALDANE, J. B. S. (1932). *The Causes of Evolution.* London.
107a. HARTERT, E. (1910). *Novit. Zool.* **17**, 484.
108. HEINROTH, O., and HEINROTH, M. (1924–33). *Die Vögel Mitteleuropas.* Berlin.
109. HESS, C. V. (1914). *Lichtsinn und Farbensinn in der Tierreihe.* Berlin.
110. HINDWOOD, K. A. (1948). *Emu*, **47**, 389.
111. HINGSTON, R. W. G. (1933). *The Meaning of Animal Colour and Adornment.* London.
112. HIRST, A. (1940). *Vict. Nat.* **57**, 133.
113. —— (1944). *Avic. Mag.*, 5th ser. **9**, 47.
114. —— (1944). Ibid. 132.
115. —— Personal communication.
116. HÖHN, E. O. (1947). *Proc. Zool. Soc. Lond.* A, **117**, 281.
117. HOPKINS, N. (1948). *Emu*, **47**, 331.
118. —— (1953). Ibid. **53**, 279.
119. —— Personal communication.
120. HUMPHRIES, C. P. (1947). *Emu*, **47**, 130.
121. HUXLEY, J. S. (1934). *Proc. Orn. Congr. Oxford*, 430.
122. —— (1944). *Evolution: The Modern Synthesis*, 3rd ed. London.
123. HYEM, E. L. (1937). *Emu*, **36**, 262.
124. IREDALE, T. Communication to CHAFFER, N., 43.
125. —— (1948). *Austr. Zool.* **11**, 163.
126. —— (1950). *Birds of Paradise and Bower-Birds.* Melbourne.
127. JACKSON, S. W. (1909). *Emu*, **8**, 233.
128. —— (1910). Ibid. **10**, 81.
129. —— (1912). Ibid. **12**, 65.
130. —— (1920). Ibid. **19**, 258.
131. —— Communication to CAMPBELL, A. J., 36.

REFERENCES

132. JARDINE, W., and SELBY, P. J. (1830). *Illustr. Orn.* **2**, 103.
133. JARDINE, B. Cited by NORTH, A. J., 205.
133a. JARMAN, H. (1953). *Emu*, **53**, 183.
134. JERRARD, C. H. H. (1925). *N.Z.L. Quart. Mag.*, June, 31.
135. KEARTLAND, G. A. Cited by NORTH, A. J., 205.
135a. KEAST, J. A., and MARSHALL, A. J. (1954). *Proc. Zool. Soc. Lond.* **124**, (3). In press.
136. KILGOUR, J. F. (1904). *Emu*, **4**, 37. Also cited by MATHEWS, G. M., 185.
137. KIRKPATRICK, C. M., and LEOPOLD, A. C. (1952). *Science*, **116**, 280.
138. KLUIJVER, H. N. (1933). *Versl. en Meded. Plantenziektenk. Dienst. Wageningen*, **69**, 1.
139. KRETSCHMER, E. (1952). *A Textbook of Medical Psychology*. London.
140. LACK, D. (1946). *The Life of the Robin*. Rev. ed., London.
141. —— and SILVA, E. T. (1949). *Ibis*, **91**, 64.
142. LAHR, E. L., and RIDDLE, O. (1944). *Endocrinol.* **35**, 261.
143. LEA, A. M., and GRAY, J. T. (1935). *Emu*, **34**, 275 (and subsequent issues).
144. LE SOUEF, A. A. C. Cited by CAMPBELL, A. J., 36.
145. LE SOUEF, A. S. (1913). *Emu*, **12**, 190.
146. LESSON, R. (1925). *Oiseau de Paradis*. Paris.
147. LEVICK, G. M. (1914). *Antarctic Penguins: A Study of their Habits*.
148. LEWIN, J. W. (1808). *The Birds of New Holland*. London.
149. —— (1822). *Natural History of the Birds of New South Wales*. London.
150. LINNÉ, C. (1758). *Systema Natura*, 10th ed., 108. Uppsala.
151. LØPPENTHIN, B. (1932). *Medd. øm Grønland*, **91**, 1.
152. LORD, E. A. R. (1939). *Emu*, **39**, 124.
153. LORENZ, K. (1935). *Journ. f. orn.* **83**, 137; 289.
154. LUMHOLTZ, C. (1889). *Among Cannibals*. London.
155. MACGILLIVRAY, J. (1852). *Narrative of the Voyage of H.M.S. Rattlesnake*, 1. London.
156. MAKIN, D. Personal communication.
157. MANDER JONES, P. Personal communication.
158. MARGERY, I. D. (1926). *Quart. J. Roy. Met. Soc.* **52**, 27.
159. MARSHALL, A. J. (1931). *Austr. Zoologist*, **6**, 355.
160. —— (1932). *Emu*, **32**, 33.
161. —— (1934). *Ibid.* **34**, 57.
162. —— (1942). *Display and Bower-building in Bower-birds*. Thesis. (Fisher Library, University of Sydney.)
163. —— (1944). *Nature*, **153**, 685.
164. —— (1947). *Proc. Linn. Soc. Lond.* **159**, 103.
165. —— (1949). *Proc. Zool. Soc. Lond.* **119**, 711.
166. —— (1949). *Quart. J. Micr. Sc.* **90**, 265.
167. —— (1950). *Emu*, **50**, 5.
168. —— (1950). *Nature*, **165**, 388.
169. —— (1950). *Ibid.* **166**, 204.
170. —— (1951). *Proc. Zool. Soc. Lond.* **120**, 749.
171. —— (1951). *Wilson Bull.* **63**, 238.
172. —— (1951). *Emu*, **50**, 267.
173. —— (1951). *Discovery*, **12**, 146.
174. —— (1952). *Ibis*, **94**, 310.
175. —— (1952). *Proc. Zool. Soc. Lond.* **122**, 239.
176. —— (1952). *Ibid.* **121**, 727.

177. MARSHALL, A. J. (1954). *Proc. Zool. Soc. Lond.* **124**, 219.
178. —— (1954). *Biol. Rev.* **29** (1), 1.
179. —— Unpublished.
180. —— and COOMBS, C. J. F. (1952). *Nature*, **169**, 261.
181. —— —— Unpublished.
182. —— and HARRISSON, T. H. (1941). *Emu*, **40**, 310.
183. MARSHALL, F. H. A. (1936). Croonian Lecture. *Phil. Trans. Roy. Soc.*, Ser. B, **226**, 423.
184. —— (1937). *Proc. Roy. Soc.* B, **122**, 413.
185. MATHEWS, G. M. (1926). *The Birds of Australia*, **12**. London.
186. MAYER, F. SHAW. Cited by IREDALE, T., 126.
186a. —— Personal communication.
187. MAYR, E. (1932). *Nat. Hist.* **32**, 83.
188. —— (1941). *List of New Guinea Birds.* New York.
189. —— Personal communication.
190. —— and GILLIARD, E. T. (1950). *Amer. Mus. Nov.* **1473**, 1.
191. —— and JENNINGS, K. (1952). Ibid. **1602**, 1.
192. McGILP, J. N. (1931). *S.A. Ornith.* **11**, 98.
193. McLENNAN, W. Cited by CAMPBELL, A. J., 36.
194. MILLER, A. H. (1949). *Science*, **109**, 546.
195. MILLER, R. S. (1937). *Emu*, **37**, 73.
196. MIYAZAKI, H. (1934). *Sci. Rep. Tohuku Imp. Univ.*, 4th ser. Biol. **9**, 183.
197. MOREAU, R. E., WILK, A. L., and ROWAN, W. (1947). *Proc. Zool. Soc. Lond.*, **117**, 345.
198. MORLEY, A. (1943). *Ibis*, 132.
199. MORRISON-SCOTT, T. C. S. (1937). *Proc. Zool. Soc. Lond.* A, 41.
200. MURRAY, R. J. Personal communication.
201. MUSGRAVE, A. Personal communication.
202. NALBANDOV, K. W. (1945). *Endocrinol.* **36**, 251.
203. NEUNZIG, K. (1921). *Fremdländische Stubenvogel.* Magdeburg.
204. NEWTON, A. (1899). *A Dictionary of Birds.* London.
205. NORTH, A. J. (1904). *Nests and Eggs of Birds found Breeding in Australia and Tasmania.* Sydney.
206. —— (1908). *Proc. Linn. Soc. N.S.W.* **33**, 799. Also ANON. (1909). *Emu*, **8**, 225.
207. NUBLING, E. (1921). *Emu*, **21**, 11.
208. —— (1939). Ibid. **39**, 22.
209. —— (1941). *Austr. Zoologist*, **10**, 95.
210. OGILVIE-GRANT, W. R. (1915). *Ibis (Jub. Suppl.)*, **2**, 31.
211. OLIVE, E. Cited by NORTH, A. J., 205.
212. O'REILLY, B. Cited by CHISHOLM, A. H. (1951). *Emu*, **51**, 75.
213. ORR, R. T. (1945). *Condor*, **47**, 117.
214. PARSONS, J. H. (1924). *Colour Vision.* London.
215. PAYKULL, G. (1815). *Nov. Act. Reg. S. Sc.*, Uppsala, **7**, 283.
216. PHILLIPPS, R. (1901). *Avic. Mag.* **7**, (7), 138.
217. —— (1905–6). Ibid., N.S. **4**, 51, 88, and 123.
218. —— (1907). Ibid. **5**, N.S. (5), 142.
219. —— (1907). Ibid., 2nd ser., **5**, (2), 54, and 3rd ser., **2**, (12), 365.
220. PLATH, K. (1944). Ibid., 5th ser., **9**, 198.
221. PLOMLEY, K. F. (1935). *Emu*, **34**, 199.
222. POULSEN, H. (1951). *Proc. Xth Int. Ornith. Congress II*, **7**, 381, Uppsala.

REFERENCES

223. POULTON, E. B. (1890). *The Colours of Animals, Their Meanings and Use.* London.
223a. RAMSAY, E. P. (1867). *Ibis*, 456.
224. ——— (1875). *Proc. Zool. Soc. Lond.* 591.
225. RAMSAY, J. S. P. (1919). *Emu*, **19**, 6.
226. ——— Cited by CHAFFER, N., 42.
227. ——— Personal communication.
228. RAND, A. L. (1940). *Amer. Mus. Nov.* **1072**, 9.
229. ——— (1942). *Bull. Am. Mus. N.H.* **79**, 351.
230. REICHENOW, A. (1897). *Orn. monats.* **5**, 24.
231. RIDGWAY, R. (1912). *Color Standards and Color Nomenclature.* Washington.
232. RILEY, G. M. (1936). *Proc. Soc. Exp. Biol. and Med.* **34**, 331.
233. ——— (1937). *Anat. Rec.* **67**, 327.
234. RIPLEY, D. (1947). *Trail of the Money Bird.* London.
235. ROAF, H. E. (1929). *Proc. Roy. Soc. B*, **105**, 371.
236. ROBERTS, N. L. (1937). *Emu*, **37**, 48.
237. ROBINSON, A. (1933). *Ibid.* **33**, 95.
238. ——— (1936). *Ibid.* 229.
238a. ROBINSON, A. M. Personal communication.
239. ROGERS, F. T. (1922). *Journ. Comp. Neurol.* **35**, 61.
240. ROGERS, J. P. Notes included by HALL, R. (1902), *Emu*, **1**, 87.
241. ——— Cited by NORTH, A. J., 205.
241a. ROMANES, G. J. (1892). 'Darwin and After Darwin. I.' *The Darwinian Theory.* London.
242. ROTHSCHILD, LD. (1895). *Novit. Zool.* **2**, 480.
243. ROWAN, W. (1925). *Nature*, **115**, 494.
244. ——— (1929). *Proc. Boston Soc. Nat. Hist.* **39**, 151.
245. ——— (1932). *Proc. Nat. Acad. Sc. Wash.* **18**, 639.
246. ——— (1938). *Biol. Rev.* **13**, 374.
246a. SALMON, H. (1953). *Emu*, **53**, 263.
247. SALVADORI, I. *Ann. Mus. Civ. Genova*, (2), **14**, 151.
248. SCHAANING, H. T. L. (1916). *Dansk Ornith. Forenings Tidss.* **10**, 145.
249. SCHLEGEL, H. (1871). *Ned. Tijdschr. v. a. Dist.* **5**, 51.
249a. SCHÖNWETTER, M. (1944). *Beitr. Fortpfl.-biol. Vögel*, **20**, 1.
250. SCHULTZ, M. (1866). *Arch. f. mikr. Anat.* **2**, 255.
251. SEDGWICK, E. H. (1946). *Emu*, **45**, 300.
252. SELVAGE, J. J. Personal communication. See also *N.Q. Bird Notes* (1948), **16**, 2.
253. SERVENTY, D. L. (1946). *W. Austr. Bird Notes*, **4**, 10.
254. ——— and WHITTELL, H. M. (1951). *Birds of Western Australia.* Perth.
255. SHARPE, R. B. (1884). *Journ. Linn. Soc. Lond.* **17**, 408.
256. ——— (1891–8). *Monograph of the Paradiseidae and Ptilonorhynchidae*, II. London.
257. SHARP, G. Cited by CHISHOLM, A. H., 46.
258. SHILLING, D. (1948). *Emu*, **48**, 64.
259. SIMSON, C. C. (1907). *Ibis*, **1** (9th ser.), 380.
260. SLUITER, J. W., and VAN OORDT, G. J. (1947). *Quart. J. Micr. Sc.* **88**, 135.
261. ——— ——— (1949). *Ibid.* **90**, 1.
262. SMITH, A. F. (1906). *Emu*, **5**, 211.

263. SODERBERG, R. (1929). *Verh. VI, Int. Orn. Kongr. Kopenhagen*, 297. Berlin.
264. STIEVE, H. (1950). *Die Naturwissenschaften*, **37**, 8 and 33.
265. STOKES, J. L. (1846). *Discoveries in Australia*. London.
266. STONOR, C. R. (1937). *Proc. Zool. Soc. Lond.* B, **122**, 475.
267. STORR, G. M. (1951). *Emu*, **50**, 184.
268. STRESEMANN, E. Personal communication.
268a. —— (1930). *Novit. Zool.*, **36**, 6.
268b. —— (1953). *Die Vogelwarte*, **16** (4), 148.
269. SULLIVAN, C. (1930). *Emu*, **30**, 110.
270. —— (1931). *Emu*, **31**, 124.
271. SUMMERS-SMITH, D. Personal communication.
272. TEMMINCK, C. J., and LAUGIER, M. (1838). *Nouv. Rec. de Planches Col. d'Ois.* **2**, pl. 575. Paris.
273. THOMSON, D. F. (1935). *Birds of Cape York Peninsula*. Melbourne.
274. TINBERGEN, N. (1951). *The Study of Instinct*. Oxford.
275. TUBB, J. A. (1945). *Emu*, **44**, 249.
276. VIEILLOT, L. J. P. (1816). *Nouv. Dict. d'Hist. Nat.* **6**, 569.
277. WADE, E. W. (1910). Quoted in *Brit. Birds*, (3), **9**, 309.
278. WAGNER, H. O., and STRESEMANN, E. (1950). *Zool. Jahr.* (Syst.), **79**, 273.
279. WALLACE, A. R. (1869). *Malay Archipelago*, **2**, 2nd ed. London.
279a. WALLER, E. Cited by COXEN, C. (1864). *Q'ld. Philos. Soc.* 23 May. Also cited by GOULD, J., 104.
280. WEISKE, E. (1902). *Orn. Monatsschr.* **27**, 41.
281. WHITLOCK, H. L. (1910). *Emu*, **9**, 181.
282. —— (1923). Ibid. **22**, 259.
283. WHITLOCK, F. L. (1924). Ibid. **23**, 248.
284. —— (1925). Ibid. **25**, 69.
285. WHITTELL, H. M. (1942). Ibid. **42**, 56.
286. WILSON, A. H. (1922). Ibid. **21**, 273.
287. WITHERBY, H. F., et al. (1948). *The Handbook of British Birds*. London.
288. WITSCHI, E. (1935). *Wilson Bull.* **67**, 177.
289. WOOD, C. (1907). *Ophthalmol.* April.
290. YEATES, N. T. M. (1949). *Journ. Agric. Sci.* **39**, 1.

SCIENTIFIC NAMES OF ANIMALS MENTIONED IN TEXT

Badger	*Meles meles*
Bat, Giant Fruit	*Pteropus giganteus*
Bird-of-Paradise, Rifle	*Ptiloris* spp.
Blackbird	*Turdus merula*
Blackcock (Grouse)	*Lyrurus tetrix*
Butcher-bird, Grey	*Cracticus torquatus*
'Cats', Marsupial	*Dasyurus* spp.
Chaffinch	*Fringilla coelebs*
Cormorant, Little Pied	*Microcarbo melanoleucus*
Cuckoo, European	*Cuculus canorus*
Cuckoo, Koel	*Eudynamys orientalis*
Crow, American	*Corvus brachyrynchos*
Crow, Hooded	*Corvus cornix*
Deer, Roe	*Capreolus capreolus*
Dingo	*Canis dingo*
Dove, Peaceful	*Ceopelia placida*
Dove, Stock	*C. oenas*
Eagle, Whistling	*Haliaster sphenurus*
Egret	*Egretta* spp.
Emu	*Dromaius novae-hollandiae*
Ermine	*Mustela erminea*
Finch, Plum-headed	*Aidemosyne modesta*
Finch, Cuban	*Tiaris* (= *Euethia*) *canora*
Finch, Junco	*Junco hymenalis*
Finch, Nonpareil	*Erythrura prasina*
Finch, Woodpecker	*Camarhynchus pallidus*
Fulmar	*Fulmarus glacialis*
Honeyeater, New Holland	*Meliornis novae-hollandiae*
Leatherhead	*Philemon* spp.
Lemming	*Lemmus lemmus*
Logrunner	*Orthonyx* spp.
Lyrebird	*Menura* spp.
Mallard	*Anas platyrynchos*
Mistletoe-bird	*Dicaeum hirundinaceum*

Nightingale	*Lucinia megarhynchus*
Penguin, Adelie	*Pygoscelis adeliae*
Penguin, Emperor	*Aptenodytes forsteri*
Pigeon, Bronzewing	*Phaps chalcoptera*
Pigeon, Green	*Chalcophaps chrysochlora*
Pigeon, Guria	*Goura cristata*
Pigeon, Wood	*Columba palumbus*
Pitta	*Pitta* spp.
Quail, California Valley	*Lophortyx californica*
Quail, Rain	*Coturnix coromandelica*
Raven, Australian	*Corvus coronoides*
Raven, European	*C. corax*
Robin	*Erithacus rubecula*
Rook	*Corvus frugilegus*
Ruff	*Philomachus pugnax*
Sandpiper, Sharp-tailed	*Erolia acuminata*
Scrub-bird	*Atrichornis* spp.
Scrub-wren	*Sericornis* spp.
Skua, Long-tailed	*Stercorarius longicaudus*
Skua, Pomatorhine	*S. pomarinus*
Snowy Owl	*Nyctea scandiaca*
Sparrow, American Song	*Melospiza melodia*
Sparrow, Golden-crowned	*Zonotrichia coronata*
Sparrow, House	*Passer domesticus*
Sparrow, White-crowned	*Zonotrichia leucophrys*
Starling	*Sturnus vulgaris*
Tern, Sooty	*Sterna fuscata*
Thornbill, Yellow-tailed	*Acanthiza chrysorrhoa*
Tit, Blue	*Parus caeruleus*
Tit, Great	*P. major*
'Turkey', Scrub	*Alectura lathami*
Warbler, Rock	*Origma rubricata*
Whistler, Golden	*Pachycephala pectoralis*
Wood-swallow, Dusky	*Artamus cyanopterus*

INDEX

Page references in italics refer to illustration.

aborigines, Australian, 98–99, 138.
accessory sex organs, 12.
adaptive radiation, 165, 170, 178.
aggression, 35, 42, 46, 63, 68, 81, 83, 86, 137, 140, 191.
amino-acids, *see* protein food.
anterior pituitary gland, 7–*8*, 11–12, *Pl. 2*; removal effects, 12, *Pl. 3*, 193.
Archbold Expedition, 141–2, 145.
Archbold's Bower-bird, *see* Bower-bird, Gold-crested Black.
Arctic breeding cycles, 12, 19–20.
Australian Museum, Sydney, 27, 72.
autumn display in British birds, 11, 13.
Avenue-builders, *4*, 26, 72, 89, 100, 106, 109, 119, 120, 177; classification, 183–4; distribution, *27*, *74*, *110*, *173*; eggs, *184*; evolution, 149, 170, 178.

Baker, John R., 15.
Banks, Sir Joseph, 6.
bats, breeding seasons, 16, 20.
Beagle, H.M.S., 90.
Beccari, C., 121, 123, 124.
Beck, Rollo, 119.
behavioural classification, 164, 183–4.
behavioural interactions, 25.
Benoit, J., 193.
Berney, F. L., 84.
Bird-of-Paradise, Crested (*Cnemophilus macgregorii*): alleged bower-bird, 144; description and habitat, 144–*145*, *Pl. 18*; juvenile, 147; Melanesian description, 146; moult, 147; relationships, 145; search for bower, 146; sexual season, 146–7; song, 145; sub-species, 145.
Bird-of-Paradise, Greater, 120.
Bissonette, T. H. and Wadlund, A. P. R., 10.
Blood, Captain N. B., 130.
Bourke, P. A. and Austin, A. F., 137, 138, 158.
BOWER-BIRD, FAWN-BREASTED (*Chlamydera cerviniventris*): avenue-builder, *4*, 101–3, *Pl. 15*; bower display, 103; bower elevation, 102; bower functions, 105; bower song, 103; breeding season, 103; copulation with skin, 103; description, 101; display-things, 102–103, *Pl. 15*; distribution and habitat, 101, *74*; intelligence, alleged, 105; low clutch size, 105; nest and eggs, 104; ventriloquial song, 103; wet season and food, 104.
BOWER-BIRD, GOLD-CRESTED BLACK (*Archboldia papuensis*): affinities uncertain, 143; description, 141–*142*; display-ground, 143; display-things, 143; distribution, 141; food, 142; gonad state, 142; sub-species, 141.
BOWER-BIRD, GOLDEN, *see* Gardener, Queensland.
BOWER-BIRD, GREAT GREY (*Chlamydera nuchalis*): abnormal bower, 91, 170; autumnal display, 97; avenue-builder, 91, *Pl. 14*; bower orientation, 93–94; builds in tropical gardens, 98; clutch-limiting mechanism, 98; collects gold nuggets and opal, 92; description and crest structure, 90; display and sexual cycle, 95, *Pl. 13*; display at bower, 96; display in military camp, 94; display-ground, 91; display-season, 94–96; display-things, zonation, 92; distribution, 89, *74*; egg season variable, 98; feed in butcher's shop, 91; females may lack crest, 90, 99; flocking, 95, 97; food, 90–91; habitat, 89; miners' pets, 92; moult, 97; nest and eggs, 96–97; old bowers left standing, 93; *Pl. 14*; predators, 98; protected by aborigines, 98–99; rainfall influences, 97–98; song at bower, 96; steals spectacles, 91; sub-species, 89–90; testis metamorphosis, 95; vocal mimicry, 96.
BOWER-BIRD, MADANG, *see* Bower-bird, New Guinea Regent.
BOWER-BIRD, NEW GUINEA REGENT (*Sericulus bakeri*): description, 119; distribution and habitat, *110*, 119, *Pl. 18*; female and habits unknown, 119.
BOWER-BIRD, NEWTON'S, *see* Gardener, Queensland.
BOWER-BIRD, RAWNSLEY'S, 26.
BOWER-BIRD, REGENT (*Sericulus chrysocephalus*): avenue-builder, *4*, 109, 112–13, *Pl. 17*; aviary observations, 114–16; bower song, 113–14, 116; bower-building and weather, 113; colour change, 111; description, 109; display-ground, 112–13, 115; display season, 112; display-things resemble female or rivals?, 118; distribution, 109, *110*, 119; eggs, 116; flocking, 117; food, 111, 116–17; habitat, 109, 112; morphology and display, 111, 118; nesting season, 116, *Pl. 17*; painting bower, 113–14; polygamy, possible, 115–16; pursues winged insects, 117; relationships, 117–18; vocal mimicry, 114; young birds, 117.
BOWER-BIRD, SATIN (*Ptilonorhynchus violaceus*): abnormal bowers, 170; aggression, 35, 42, 44, 46, 62–63, 68, 181;

INDEX

antiquity of painting, 191–2; arboreal display, 40; autumn bowers, 30–31; avenue-builder, *4*, 29, 36, *Pls. 1, 5*; beak colour, significance, 191; blue birds' savagery, 68; bower-building, 31–32; bower-display, 37, 39, 62; bower functions, 42–44; bower orientation, 40, *41, 42–43*; bower wrecking, 44, *Pl. 1*; bower-song, 39, 66; breeding cycles, 7–12, 29, 57, 132, 146–7, *Pls. 7, 8, 9*; castration, 68; central nervous system, 35, 61, 63; clutch limiting mechanism, 71; colour change, 28–29; colour collection, 46, 64; colour collection by young, 46, 59–60; colour phases, 68; colour selection, 35, 38, 47, 48–49; colour vision, 49; copulation with skin, 55; delayed ovulation, utility, 56; description, 29; discrimination, 37–38; displaced combat, 62, 190; display, communal, 61; display without female, 67; display, flock, 30; display functions, 42–43; display, juvenile, 166; display, postnuptial, 30, 58; display-things, resemblance to female or rivals?, 48–50, 64, 190–1; display, winter, 31; distraction display, 33; distribution and habitat, 26, *27, Pl. 4*; eggs, 55; experiments, *40–41*, 42, 44–46, 47–48, 53–54, 59–60, 68, 189–91; external stimuli, 32–35, 67; female at bower, 39–40, *Pl. 5*; fetishism, so-called, 64; flock song, 30; flocking, 26, 29–30, 36, 58; food, 30, 54, 56–67, 71; green males, 37, 191; green males breeding, 43; green males in competition, 68; hatching date, 57; intelligence, alleged, 67, 185; kills captive blue finches, 61–62; learning ability, 67; longevity, 70–71; moult, 28, 49; moulted feathers used, 49; mutant, 26; nesting behaviour, 54–55, *Pl. 6*; painting bower, 65, 70; pair-bonds, 31, 67; photoperiodicity, 31; polygamy, possible, 55, 67, 69–71; predator-pressure, 71; protein food for young, 56, 67; 'recreation hypothesis', 67; refractory period, 33; sex hormones, 35, 61, 63, 68; sex reversal?, 68; sexual cycle, experimental modification, 53–54; sexual dimorphism, 67; sexual periodicity, 31; sexual resurgence, 36; sexual selection, 69; sexual synchronization, 56–66; similarity to *Chlamydera*, 86; stealing blue flowers, 46; stealing others' possessions, 44–47, *Pl. 1*; sub-bowers and platforms, *4*, 36, 69, *Pl. 6*; submissive green birds, 46, 68; sub-species, 26; temperature, 33–34, 57, 71; territory, 33, 44–46, 66, 191; territory call, 40; territory rivalry, 44–45; testis metamorphosis, 29, 30, 57, *Pl. 9*; tongue, 56; transported across equator, 33; vocal mimicry, 40; wrecking and robbing bowers, 44, 63.

BOWER-BIRD, SPOTTED (*Chlamydera maculata*): abnormal bowers, 91, 170, *171*; aggression, 81–83, 86; auditory stimuli, 80; autumn bowers, 74; avenue-builder, *4*, 74, *Pl. 10*; bower orientation, 75, 86; builds in gardens, 74; colour selection, 77–78; copulation at bower, 81; description and crest structure, 72–73, *Pls. 10, 11*; discrimination, 77, 87; display at bower, 80–81, *Pl. 11*; display, communal, 74, 80; display functions, 88; display, juvenile, 87; display-things, 75–76, 77–78, *Pl. 10*; display-things, resemblance to female or rivals, 87; display-things, zonation, 78; distraction display, 83–84; distribution and habitat, 73–74, *Pl. 10*; eats poisonous berries, 73; erectile crest, 72–73, 82–83; flocking, 85; food, 73, 84; innate behaviour pattern, 87; intelligence, alleged, 80, 186; male may display alone, 85; nest and eggs, 85; nesting season, 84; painting bower, 78; platforms, 75; polygamy, possible, 80; predators, 85; rainfall effects, 85; relationship with Satin Bower-bird, 86, 88; seasonal gonad states, 79, 82, 85; slaughtered by white men, 99; steals car ignition keys, 77; steals glass eye, 76; stray near coast, 73; sub-species, 73; territorial call, 83, 87; testis metamorphosis, 79, 85; vocal mimicry, 81, 83–84.

'BOWER-BIRD', TOOTH-BILLED, see Stagemaker.

BOWER-BIRD, YELLOW-BREASTED (*Chlamydera lauterbachi*): additional avenues, *4*, 107–8, *Pl. 16*; breeding season, 107; display-ground, 106–8; display-things, 107–8; description, 106; distribution and habitat, 74, 106, *Pl. 18*; gonad state, 107; relationships, 108.

BOWER-BIRD, YELLOW-FRONTED, see Gardener, Golden-maned.

BOWER-BIRDS: aestheticism, 185–6; behaviourisms, exaggerated, 181; continental changes and distribution, 140, 168, 172–3, 177–8; destruction by whites, 190; eggs, 171–2, 184; evolutionary rates, 181; intelligence, alleged, 1–2, 53, 67, 80, 105, 185; invested with human attributes, 185–7; learning, 67, 185–6; origin of name, 6; slang expression, 6; *see also under individual species, including* Gardeners.

bower-building: antiquity, 118, 192; classification, 125–6, 129, 143, 181–4; displacement activity, 171; evolution, 108, 165, 170–1, 192; physical actions, 39, 172; restricted to bower-birds, 182; weather factors, 33, 34, 189.

bower-form and phylogeny, 181.

bower orientation: experiments, *40–41–*42; function, 42–43; Great Grey

INDEX

Bower-bird, 93–94; Satin Bower-bird, 40–43; Spotted Bower-bird, 75, 86.
Bowers: aberrant, 91, 170, *171*, *174*; functions, 2, 42–44; homologies, 108, 126, 133; *see also under individual species*.
breeding cycles: malleability, 15; mammalian, 20–21; mechanism, 7–8–12.
breeding seasons: Arctic, 19; behavioural interactions, 25; cuckoos, 24; desert and drought, 15, 19; evolution, 15; food, 19, 22; fruit bat, 20; general, 7; Golden Whistler, 20; gonad states, 22; internal rhythm, 25; natural selection, 25; New Hebridean, 20; rainforest birds, 18; timing, 12, 14, 22–23, 25; tropical birds, 14, 18, 20, 25; variability, 24; weather factors, 12, 17–18.
British Association, 100.
British Museum (Natural History), 101, 127.
Broadbent, Kendall, 97, 135, 136, 138.
bush-fires, 65.

Campbell, A. G., 113.
Campbell, A. J., 135.
Campbell, A. J. and Barnard, H. G., 158.
Cat-bird, Green (*Ailuroedus crassirostris*): breeding season, 151; description, 148, *Pl. 24*; display, 150; distraction display, 192; distribution and habitat, *149*, *Pls. 18, 22*; flocking, 151; food, 150–1; male tends young, 151; nest and eggs, 151, *Pl. 24*; song, 148, 150; sub-species, 149–50.
Cat-bird, Tooth-billed, *see* Stagemaker.
Cat-bird, White-eared, *see* Cat-bird, White-throated.
Cat-bird, White-throated (*Ailuroedus buccoides*): description, *152*–153; distribution and habitat, *149*, 152, *Pl. 18*; eggs, 153; song, 153; sub-species, 153.
Cat-birds (Ailuroedinae): arboreal display, 167; beaks, *152*, 168, *Pl. 25*; classification, 183–4; distribution and habitat, *149*, 168, *Pls. 18, 22*; eggs, 185, *Pl. 24*; feeding habits, 168; speciation, 169.
Central nervous system, 7, 9, 12, 35, 61, 63.
Chaffer, Norman, 52, 67, 75, 81, 82, 106, 107, 108, 117, 132, 170.
Chisholm, A. H., 77, 159, 167.
Chlamydera spp.: distribution, 74, 177, 179; evolution, 177–9.
classification (of Ptilonorhynchidae), 3, 117, 122, 143, 168, 181–4, 192; behavioural, 3, 122, 177, 183–4; Iredale's, 182–3; Mathew's, 182; Stonor's, 3; unreliability of epigamic features, 181–2.
Cnemophilus macgregorii, *see* Bird-of-Paradise, Crested.
Coleman, W. H., 162.

colour display, *see under various bowerbirds*.
colour phases, *see under various bowerbirds*.
colour selection, *see under various bowerbirds*.
colour vision in birds, 50.
Condon, H. T., 94.
Cornwall, E. M., 158, 162.
'*Corymbicola mestoni*', 135.
Coues, E., 154.
courtship feeding, 53.
Coxon, Charles, 27.
Crandall, Lee S., 102.
'Crop milk', 23.
Cuckoos, external stimuli, 24.

d'Alberti, M. L., 121.
Darwin, Charles, 2, 69, 90, 185.
David, T. W. Edgeworth, 180.
Day, W. S., 92, 137, 139.
daylength: breeding seasons, 24–25; bower-building, 31–34, 189–90.
De Vis, C. W., 130, 131, 135, 141, 146.
Diggles, S., 26.
displacement activity: bower-building, 171–2; combat, 62, 64, 190–1; courtship feeding, 53, 65, 172; display, 187; mimicry, 55, 83–84, 167; with flower petals or nest material, 166.
display: aphrodisiacal quality, 187; convergent evolution, 164; effect of age or infirmity, 192; evolution, 166, 170; innateness, 2, 59, 87, 166; similarity in avenue-builders, 177; taxonomic factor, 168.
display-grounds: heighten reproductive isolation, 165; in unrelated groups, 182; specificity in bower-birds, 165.
display-things: discrimination, 6, 29, 37–38, 48–50, 60, 77, 87, 92, 170, 190–1; resemblance to female or rivals?, 48–50, 64, 87, 105, 172, 178, 190–1; selected compulsively, 185.
distraction display: Green Cat-bird, 192; Satin Bower-bird, 55; Spotted Bowerbird, 83–84; vocal mimicry, 55, 83–84, 167.
distribution: Avenue-builders, *173*; Cat-birds (incl. Stagemaker), *149*; *Chlamydera*, spp., 74; Maypole-builders, *174*; *Ptilonorhynchus*, 27; *Sericulus* (= *Xanthomelus*) spp., *110*.
drought, effects, 19, 73, 84, 98.

Edwardian prose, of collector, 71.
eggs and classification, 86, 116, *184*, 192.
Elton, Charles, 19.
evolution: adaptive radiation, 165, 170, 178; bower-birds, 169, 178; bower-building, 165, 170, 192; breeding seasons, 15, 25; Cat-birds, 168–9; convergence in display, 164–5, 182.

experimental findings, 37, 44–46, 53–54, 59–60, 189–91.
external inhibitors, 12, 17–19.
external stimuli, 7–8–9, 14, 17–18, 165; Bower-birds, 31–35, 57, 84, 97; Cuckoos, 24; general, 7, 19, 22, 24, 84, 187, 189–90.

fetishism, so-called, 64.
Flinders, Matthew, 6.
flocking, *see under separate species*.
floral changes and distribution, 27, 140, 168–9, 172, 177–8.
Fowler, H. W., misleading comparison, 6.
Freud, Sigmund, 65.

Gannon, G. R., 51.
Gardener, Brown (*Amblyornis inornata*): builds maypole-bower, 5, 123, 125–6, Pls. *19*, *20*; distribution, 123, *174*; flower collection, 124; garden and display-things, 124–5, Pls. *19*, *20*; habitat, 125, Pl. *18*; hut of orchid stems, 124; vocal mimicry, 124.
Gardener, Golden-maned (*Amblyornis flavifrons*): bower, behaviour and female undescribed, 134; classification, 183; description, *134*.
Gardener, Orange-crested Striped (*Amblyornis subalaris*): atypical bower, *128*; builds maypole bower, 5, 128–9, Pl. *20*; description, 127; distribution, 127, 129; garden and display-things, 128–9, Pl. *20*; glue, alleged use, 128–9; habitat, 127, Pl. *18*; nest and egg, 129.
Gardener, Queensland (*Prionodura newtoniana*): aggression, 137, 140; builds maypole-bower, 5, 136, Pl. *23*; breeding season, 139–40; description, *8*, 135, 136; display-stick and display, 136–8, Pl. *23*; distribution and habitat, 136, 140, Pl. *22*; enormous bower, 136; flower collection, 137; food, 140; garden and display-things, 136–9; hut-like bower, 138; nest and eggs, 139; relationship with *Amblyornis*, 140; sub-bowers, 138; vocal mimicry and song, 138.
Gardener, Yellow-crested (*Amblyornis macgregoriae*): bower homologies, 133; breeding season, 132; builds maypole-bower, 5, 130–3, Pl. *21*; description, 130, *131*; distribution and habitat, 130, Pl. *18*; food, 133; gonad states, 132; nest and egg, 133; song, 133; sub-species, 130.
Gaukrodger, D. W., 75, 78, 80, 83, 84, 87.
George IV, 109.
Gibbs, G., 85.
Gilbert, John, 89.
Gilbert, P. A., 51, 116.
Gilliard, E. T., 106, 141, 142.
Goanna (*Varanus* spp.), 71, 98.
Goddard, M. T., 113.

Golden-bird (*Sericulus aureus*): description, *120–121*; distribution and habitat, 121, Pl. *18*; relationships, 122; reputed bower, 122; sub-species, 121–2; vestigial structure, 122.
'Golden-bird', Crested, *see* Bird-of-Paradise, Crested.
gonadotrophins, 9, 12, 193.
Goodwin, A. P., 144, 128, 130, 132.
Gould, John, 2, 6, 27, 72, 75, 90, 100, 109, 123, 148.
Grant, R., 162.
Green, B., 162.
Gregory, A. C., 26.
Gwynne, A. J., 150.

Haldane, J. B. S., 70.
Hingston's hypothesis, 63.
Hirst, A., 28, 42, 46, 57, 59, 68, 116.
Hopkins, Nancy, 92, 95.
hunger, 17, 187.
Hunstein, C., 127.
Huxley, Julian S., 62, 63, 68, 166.
Huxley, T. H., 100.
hybridization, 166.
hypophysectomy, 12.
hypophysis, *see* anterior pituitary.
hypothalamus, 7, 8–9, 187, Pl. *2*, 193.

innate behaviour: Satin Bower-bird, 59; Spotted Bower-bird, 87.
intelligence, supposed: Fawn-breasted Bower-bird, 105; general, 1–2, 67, 185; Satin Bower-bird, 52–53, 67; Spotted Bower-bird, 186.
internal rhythm, 7, 14, 21, 25, 193.
interstitial cell (interstitium): constitution and cycle, 8–11, Pls. *2–3*; seasonal exhaustion, 10–11, Pl. *3*; tumours in Man, 187.
Iredale, Tom, 6, 107, 122, 130, 146.
isolation, behavioural, 169.

Jackson, S. W., 71, 75, 77, 112, 137, 154, 156, 158, 161, 162, 163.
Jan Mayen Expedition, 19.
Jardine, W. and Selby, P. J., 90.
Jardines of Cape York, 101, 103.
Jarman, H., 78, 80.
Jennings, K., *see* Mayr, E.
Jerrard, C. H. H., 80, 82.
Johnstone, Inspector, 155.

Keartland, G. A., 97.
Keast, J. A., 85.
Kokoda Track, 130.

Lamarckism, 53.
Lauterbach's Bower-bird, *see* Bower-bird, Yellow-breasted.
learning in Bower-birds, 67, 186.
lemmings, 19.

INDEX

Lesson, R., 120, 121.
Levick, G. M., 15.
Lewin, John, 109.
Leydig cells, *see* interstitial cell.
Light, *see* photoperiodicity and daylength.
Linné, C., 120.
Linnean Society of London, 90.
Loke Wan Tho, 142.
Loria loriae, 146.
Lumholtz, C., 159, 162, 163.

MacGillivray, John, 100–1.
MacGillivray, Wm., 102–3.
McGilp, J. N., 76.
Macgregor Expedition, 128, 130.
Macgregor, Lady, 130.
Macgregor, Sir William, 144.
Macleay, Sir William, 90.
McLennan, W. M., 76, 84.
Macquarie, Governor Lachlan, 6.
Madang Bower-bird, *see* Bower-bird, New Guinea Regent.
Mammalia: autumn activity, 16; intra-uterine development, 15.
Man: adrenal cortex tumours, 187; behaviour, 187; cerebral cortex, 187; critical and moral faculties, 187; displacement activity, 187; hypothalamic lesions, 187; 'instinctive' drives, 188; Leydig cell tumours, 187; menstrual cycle, 21; music-making, 187; psychogenic influences, 187; sex hormones and display, 187.
Marshall, F. H. A., 11.
Marsham Records, 24.
Mathews, Gregory M., 117.
Mayer, F. Shaw, 107, 108, 122, 132, 133, 141, 142, 146, 147.
Mayr, Ernst, 28, 124, 144, 153.
Mayr, E. and Gilliard, E. T., 142.
Mayr, E. and Jennings, K., 89, 111, 118.
Maypole-bowers, 5, *Pls. 19–21, 23*; aberrant, *175–177*; evolution, 126, 173–4; fundamental features, 5, 125–6, 129, 133, 173; protective function of cone, 174, 176.
Maypole-builders, 123, 126, 130, 135; classification, 183, 184; distribution, 172, 174; eggs, *184*; evolution, 169–70.
Melanesian: description of *Cnemophilus*, 146; legend, 105; taxidermy, 120.
menstrual rhythm, 21.
Mesozoic reptiles, 70.
Meston, A., 135.
migration, bird: display and synchronization, 21; timing, 12.
Miller, R. S., 151.
mimicry, vocal, *see* vocal mimicry.
Miyazaki, H., 16.
Morley, A., 13.
Morrison-Scott, T. C. S., 47, 50, 52.
moult and gonad cycle, 11, 13.
Mueller, Baron F. J. H. von, 127.
Muit, 16.

natural selection, *see* Evolution.
nest-building, bi-sexual nature, 171.
nesting behaviour, 55.
New Hebridean Expedition, 20.
New Holland, 6, 26, 72.
Newton, Alfred, 6, 135.
non-breeding in birds: Arctic, 19–22; desert, 20, 22.
North, A. J., 74, 83, 159, 161, 162, 170.]
Nubling, E., 31, 48, 51.

Olive, E., 97, 136.
O'Reilly, B., 114.
orientation, *see* bower orientation.

Paedogenesis, 178.
painting of bower: antiquity, 65, 117, 191–2; inconstancy, 51–52; materials used, 51–52, 78, 114; origin and function, 53; physical actions in, 172; Regent Bower-bird, 115–16; reproduction without, 70; Satin Bower-bird, 52, 70; Spotted Bower-bird, 78; sub-adults, 52.
pair-bond, 166–7.
Paykull, G., 149.
penguins, Antarctic, 15, 25.
Phillipps, R., 111, 114–16.
photoperiodicity (and stimulation), 16–20, 25, 31–32.
phylogeny, *see under* evolution, and classification.
Pigeon, Green, 109.
plume-trading era, 119, 120.
poisonous fruits, eaten by birds, 73.
polygamy, possible: Regent Bower-bird, 115–16; Satin Bower-bird, 55, 58, 67, 69–71; Spotted Bower-bird, 80.
predators, 167, 176.
Prince Regent, 109, 110.
prolactin, 8–9, 11, 58.
protein food for young, 22–24, 56, 67, 104, 116–17, 140, 151, 162.

Queensland Museum, 135.

rainfall and reproduction, 18, 85, 96–98, 162.
Ramsay, E. P., 109, 112, 154, 163.
Ramsay, J. S. P., 67, 81.
Rand, A. L., 103, 104, 105, 142, 151, 152.
Rattlesnake, H.M.S., 100.
'recreation hypothesis', 1–2, 80, 90, 185.
refractory period, 9–10, 13, 18, 33, 193.
reproductive isolation, 165.
rhododendron, white-flowering, 127; migration, 180.
Ripley, Dillon, 124, 125.
Robinson, Angus, 75, 79, 84.
Rogers, J. P., 96, 97.
Romanes, G. J., 185.
Rowan, William, 16.

INDEX

Salvadori, I., 122, 146.
Satin-bird, *see* Bower-bird, Satin.
Schlegel, H., 123, 124.
Scotswoman, shipwrecked, 100.
Scottowe MS., 110.
Sedgwick, E. H., 93, 94, 96.
Selvage, J. J., 93, 95.
Sericulus spp.: distribution, *110*, 180; evolution, 179–80; relationships, 180; some bowers unknown, 180.
Serventy, D. L., and Whittell, H. M., 73.
sex hormones, 2, 9, 12, 61–62, 65, 68.
sex reversal, possible, 68.
sexual cycle, experimental modification, 53–54.
sexual drive, dynamic metamorphosis, 65.
sexual fetishism, so-called, 66.
sexual selection, 69–70.
sexual sublimation, 65–66.
Sharp, G., 138.
Sharpe, R. Bowdler, 122, 127.
Shelley, P. B., 6.
sign stimuli, possible, 56, 63, 66, 84, 191.
Simson, C. C., 128, 132.
Sluiter, J. W. and van Oordt, G. J., 11.
Soderberg's Theory, 171–2.
song, nocturnal bird, 163.
Sooty Tern, 14, 20.
speciation: Cat-birds, 169; Spotted Bower-birds, 73.
species refuges, 178, 180.
Stagemaker (*Scenopoeetes dentirostris*): arboreal display, 158; beak, 154, 163, 168–9, *Pl. 25*; concealing plumage, 159, 168; description and habitat, 154, *Pl. 22*; display functions, 163–4; display-ground and display, 168, *Pl. 25*; display season and gonad states, 155, 161, 164, *Pl. 26*; distribution, *149*, 154; experiment with dead bird, 164; external stimuli, 155; feather re-arrangement during song, 159–*160*; food, 154, 162; lack of 'intelligence', 186; leaf selection and collection, 155–7; nest and eggs, 162; nesting season, 161–2; regurgitates fruit stones, 154; relationships, 154; sexual and display cycles parallel, 163; singing-stick, 154–5, 158; uses 'anvil' to break shells, 154, *Pl. 25*; vocal mimicry and advertisement, 158–9, 163, 167–8; young, 162.
Stieve, H., 15.
Stonor, C. R., 3.

Stresemann, E., 144, 166, 192.
Sullivan, C., 75.
sympatho-adrenal system, 63.
temperature influences: breeding, 16–17, 57, 71; display, 33, 34, 155, 189.
territory, 2, 30, 33, 44–46, 61–64, 66, 70, 80–82, 94, 140, 159, 163, 165, 167–8, 180, 191.
testes, aberrant non-breeding, 86.
testis cycle outlined, 10–14, *Pls. 2–3*.
testis, metamorphosis, 10–12, 14, 20, *Pl. 3*; Great Grey Bower-bird, 95; Satin Bower-bird, 29–30, 57, *Pl. 9*; Spotted Bower-bird, 79, 85, *Pl. 13*; Stagemaker, 161, *Pl. 26*.
Thomson, Donald, 102.

ventriloquial powers, 103, 116.
vestigial feather structure, 122.
Vieillot, L. J. P., 149.
vivipary, in unrelated groups, 182.
vocal mimicry: biological value, 163; Brown Gardener, 124; Chisholm's classification, 167; displacement activity and distraction display, 55, 83–84, 167; evolution, 163; Fawn-breasted Bower-bird, 100; functions, 163, 166–7; Great Grey Bower-bird, 96; inheritance, 166–7; rain-forest birds, 167; Regent Bower-bird, 114; Satin Bower-bird, 40; Spotted Bower-bird, 81, 83–84; Stagemaker, 158–9, 163, 167–8.

Wagner, H. O., and Stresemann, E., 19.
'wakefulness theory', 16.
Wallace, Alfred Russell, 121, 149.
Waller, Eli, 109, 113.
weather effects, 12–13, 15, 17–18, 22, 24, 33–35, 57, 85–86, 97–98, 189.
Weiske, Emil, 105, 129, 132, 133, 153.
Whitlock, F. L., 75, 76, 79, 81, 82, 84, 91.

Xanthomelus (= *Sericulus*), *110*, 119–22.
'*Xanthomelus macgregorii*' (*see* Bird-of-Paradise, Crested).

Yeates, N. T. M., 16.
Yogai, 16.
Young birds: need special food, 22–24, 56, 67; significance of size, 22–23.

Zoological Society of London, 6, 72.

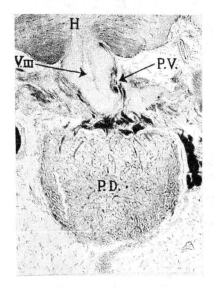

a. Brain and anterior pituitary gland. H = hypothalamus; V.III = third ventricle of brain; P.V. = hypophysial portal vessels; P.D. = pars distalis of anterior pituitary.

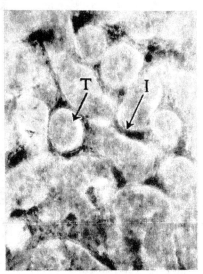

b. Testis of young bird. T = inactive seminiferous tubule; I = tract of juvenile interstitial (Leydig) cells which as yet contain only faint traces of cholesterol.

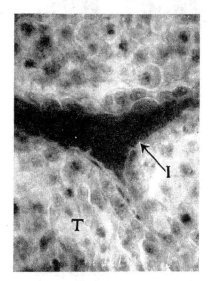

c. Testis of adult during spermatogenesis. T = tubule showing spermatocytes; I = actively secreting cholesterol-positive interstitium (producing sex hormone).

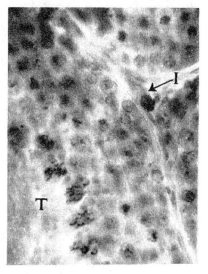

d. T = tubule showing bunched spermatozoa; I = restricted tract of interstitium showing a pale fuchsinophil cell and a dark, still heavily lipoidal cell. The interstitium is approaching exhaustion.

PLATE 2. PART OF THE INTERNAL MECHANISM OF DISPLAY AND REPRODUCTION IN PASSERINE BIRDS (p. 10)

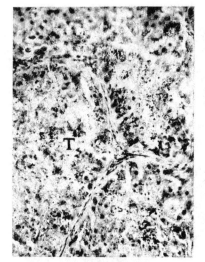

e. The interstitium is exhausted and the tubules (T) are necrosing. The black mottling is the initial appearance of the post-nuptial lipids which form as the tubules collapse. Overall, the metamorphosing organ becomes smaller and 're-gressed'.

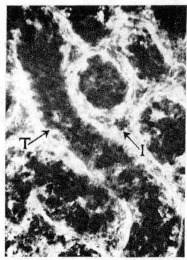

f. The tubules (T) are small and packed with lipids that have in turn become a vehicle for cholesterol. The new interstitium (I) is developing in readiness for the next sexual season. It is suspected that the metamorphosed tubules are of the nature of a temporary endocrine gland.

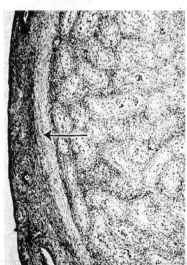

g. Low-power picture of phase shown in *f*. The tubules appear white, not dark, because the lipids have been dissolved away during the employment of a technique that shows the new testis-wall (arrowed) developing inside the old one.

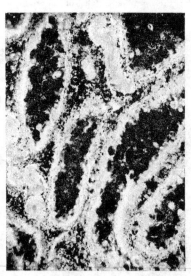

h. For comparison with *f*. The removal of the anterior pituitary gland and consequent elimination of gonadotrophic hormones results in a tubule metamorphosis and interstitial regeneration similar to that which occurs naturally at the end of the breeding season.

PLATE 3. PART OF THE INTERNAL MECHANISM OF DISPLAY AND REPRODUCTION (*continued*)

A blue male's bower is built each year in the sub-tropical undergrowth within a hundred yards of the bridge spanning Bola Creek in the Sydney National Park.

PLATE 4. HOME OF THE SATIN BOWER-BIRD (*Ptilonorhynchus violaceus*)

Blue male on the display-ground about to perform before a female within the bower.
(*Photo: Norman Chaffer*)

The male in display with a decoration in his beak. The green female is watching him.
(She is standing beneath the λ-shaped stick in the shadows.) This display-ground is not
typical in that it extends around the side of the bower (cf. Plate 1).
(*Photo: Norman Chaffer*)

PLATE 5

Primitive arena or platform built by young green male Satin Bower-birds, and frequented by numbers of birds of both sexes. The dominant blue bird of the area is a visitor.
(*Photo: P. A. Gilbert*)

Female Satin-bird at nest. She is protectively coloured and has no help from the brilliant blue male in nest-building, incubation, or the feeding of the young.
(*Photo: Norman Chaffer*)

Testes of blue male at time of bower-building. The seasonal enlargement has begun.

Testes of blue male at height of display late in September.

Testes of green male taken three weeks later in possession of former bird's bower and mate.

PLATE 7. SEASONAL GONAD STATES IN THE SATIN-BIRD

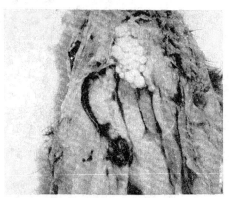

Ovary and oviduct just after beginning of male display. The seasonal modification has begun.

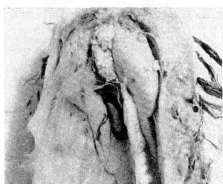

Ovary and oviduct after first flight of the young. The oviduct, through which eggs have passed, is still hypertrophied.

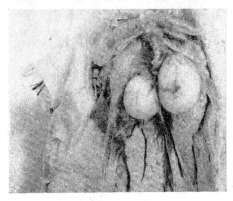

In contrast to the reduced ovary, the male organs of the same period remain productive. The displaying male has remained potentially polygamous while the female was occupied at the nest.

PLATE 8. SEASONAL GONAD STATES IN THE SATIN-BIRD

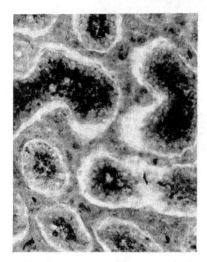

a. Heavily lipoidal tubules and juvenile lipoidal interstitium of a blue male taken in an autumn flock. At this period spasmodic visits are made to the bower territory but only snatches of display occur.

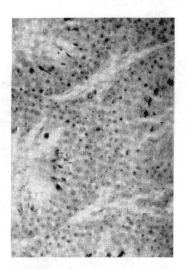

b. Bunches of sperms have appeared within a month of winter bower-building and the beginning of sustained display. (The female reproductive apparatus remains relatively unmodified—see Plate 8.)

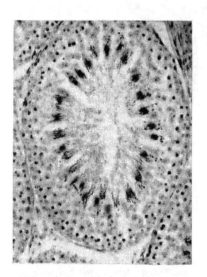

c. The tubules reach the height of productivity early in October when the male is gyrating at the bower in front of the female.

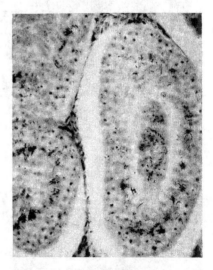

d. Post-nuptial metamorphosis is just beginning. The organs will now metamorphose to an internal appearance similar to that shown in *a* above.

PLATE 9. SEASONAL GONAD CHANGES IN THE SATIN-BIRD

Bower and bone-strewn display-ground under a wild currant bush in the hot, open savannah forest of Eastern Australia. (*Photo: S. W. Jackson*)

Male at its bower. The female stands behind during his display.
(*Photo: Norman Chaffer*)

PLATE 10. THE SPOTTED BOWER-BIRD (*Chlamydera maculata*)

Description of Plate 11

a. Erectile rose-lilac nuchal frill.
b. Differential structure of coloured erectile feathers (larger one) and the ordinary neck feathers surrounding it ($\times 5$).
c. Distal part of an individual barb of the coloured feather shown in b ($\times 350$).
d. As c, at a higher magnification to reveal structure and arrangement of the individual feather cells that alter white light and are responsible for the rose-lilac colour of the nuchal frill ($\times 600$).
e. More proximal part of an individual barb for comparison with c. At this point the specialized colour-producing cells become rarer and, as a result, only odd points of altered light appear when viewed under the microscope ($\times 350$).
f. A male in display cavorts among the bones, watched by his mate concealed behind the bower. The rose-lilac neck-frill is raised during times of heightened excitement. It was depressed when the picture was taken.

(Photo: J. S. P. Ramsay)

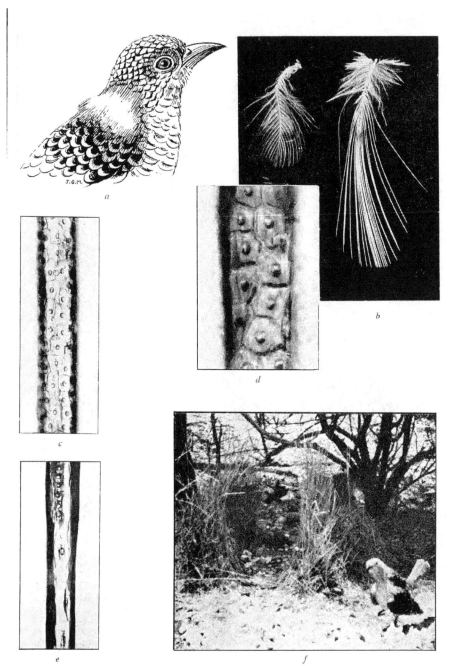

PLATE II. DISPLAY IN THE SPOTTED BOWER-BIRD

The Spotted Bower-bird lives in the open forest of inland Australia. The peculiarly marked eggs are fairly typical of those of other avenue builders.
(*Photos: S. W. Jackson and F. L. Whitlock*)

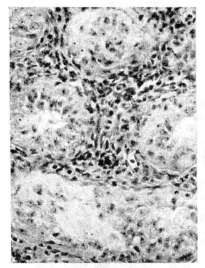

Testis of Spotted Bower-bird in autumn when display-grounds were deserted. It is in a metamorphosed condition (comparable with that of the Satin Bower-bird in Plate 9a) but it was histologically fixed in gin and so the lipids were not preserved.

Testis of Great Grey Bower-bird taken in May when bowers were in disrepair. The tubules are clear of lipids and the interstitium is heavily lipoidal—the new display-season (see illustration below) is about to begin.

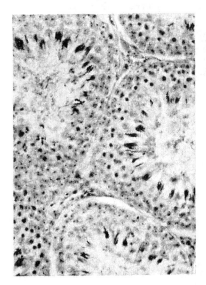
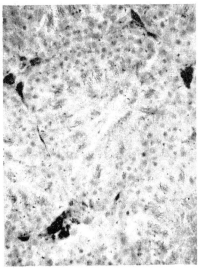

Testis of Spotted Bower-bird taken while renovating its bower in November at height of display-season.

Testis of Great Grey Bower-bird taken near freshly decorated display-ground in September. Bunched spermatozoa have appeared.

PLATE 13. SEASONAL GONAD CHANGES IN SPOTTED AND GREAT GREY BOWER-BIRDS

Freshly built bower of Great Grey Bower-bird (*C. nuchalis*) in *Melalecuca* clump on Cape York Peninsula.

Previous season's bower. The accumulation of bleached land shells has been removed to the new display-ground above.

PLATE 14

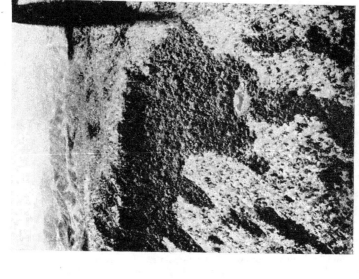

Loloki Valley, Port Moresby, is a typical home of this savannah species in New Guinea. It lives also on the north-eastern tip of Australia.

Bower of Fawn-breasted Bower-bird (*C. cerviniventris*) on raised display-ground in New Guinea. The decorations are small green fruit. (*Photo: Lee S. Crandall*)

PLATE 15

Bower of the Yellow-breasted Bower-bird (*Chlamydera lauterbachi*) of the high grasslands of New Guinea. The hand of the Melanesian indicates berries in one of the two end-avenues. The middle section is composed of two parallel walls as made by all other avenue-builders (see Plates 5, 10, 14–15, and 17). (*Photo: Norman Chaffer*)

Bower of younger, less experienced bird as shown from an end view. The end-walls are much less substantial than those in the bower above. (*Photo: F. Shaw Mayer*)

Female Regent Bower-bird (*Sericulus chrysocephalus*) of Australian rain-forests at nest with young. The brilliant black and gold male does not help incubate, or feed the young.
(*Photo: Norman Chaffer*)

Bower of Regent-bird with its few palm-seed decorations inside. Like the males of certain other avenue-builders, the owner paints the inside walls of the bower.
(*Photo: K. F. Plomley*)

Sunny *kunai*-grass, with adjacent rain-forests, is the home of the Yellow-breasted Bower-bird of New Guinea. The colours of the male harmonize with the yellow grassland.

The misty mountain rain-forests of New Guinea are the home of maypole-builders, Golden-bird, New Guinea Regent-bird, and numerous Birds-of-Paradise.
(*Photo: Australian Museum*)

The decorations on the display-ground are principally fruit and flowers. The central 'maypole' sapling and basal cone are not shown (see Plate 20). (*Photo: S. Dillon Ripley*)

PLATE 19. THE HUT OF THE BROWN GARDENER (*Amblyornis inornatus*) OF NEW GUINEA

Beccari's nineteenth-century drawing of the bower of the Brown Gardener, including the basal cone within the hut.

Edwardian artist's impression of the bower, including basal cone, of the Gold-crested Striped Gardener (*Amblyornis subalaris*).

PLATE 20

Edwardian artist's impression of bower, including basal cone, of Yellow-crested Gardener (*Amblyornis macgregoriae*).

Modern photographic reality.
(*Photo: Norman Chaffer*)

PLATE 22. RAIN-FOREST OF TROPICAL NORTH-EASTERN QUEENSLAND, THE HOME OF THE QUEENSLAND GARDENER, GREEN CAT-BIRD, AND STAGEMAKER
(*Photo: Queensland Govt.*)

Bower of the Queensland Gardener Bower-bird (*Prionodura newtoniana*). It is essentially an extended cone, decorated with orchids and mosses which continue to grow in the new situation.

Near view of the 'display-stick' beneath the decorated fabric of the bower. It is probably from here, beneath his hanging garden, that the male displays his golden plumage to the female.

Green Cat-bird (*Ailuroedus Crassirostris*) at nest with young.
(*Photo: Norman Chaffer*)

Nest and eggs of Green Cat-bird hidden in a tangle of rain-forest vines in eastern Australia.

PLATE 24

The Stagemaker or Tooth-billed Cat-bird (*Scenopoeetes dentirostris*).
(*Photo, from skin: S. W. Jackson*)

The stage. The bird grinds through the leaf petioles with its serrated beak and places fresh leaves, upside down, each morning. From a 'singing-stick'—a twig or liana—some feet above it pours an almost continuous stream of melody and vocal mimicry into the rain-forest. It sometimes breaks snail-shells on an 'anvil' beside the display-ground.
(*Photo: S. W. Jackson*)

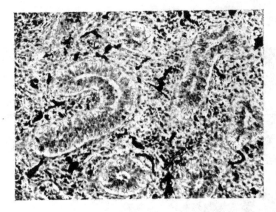

Testis in mid-August before stages were made. The tubules are narrow, and interstitial cells are small and undispersed.

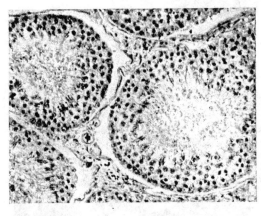

Bunched spermatozoa have appeared in November when birds are singing from above all stages in the locality.

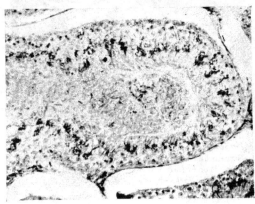

Post-nuptial metamorphosis occurs late in December when the stages are allowed to fall into disrepair and withered leaves are no longer replaced by fresh ones.

PLATE 26. SEASONAL GONAD STATES IN THE STAGEMAKER

Date Due

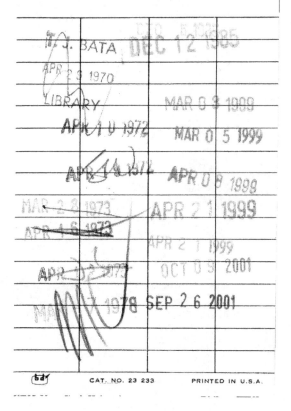

QL696 .P2M3
Marshall, Alexander James
Bower-birds

DATE	ISSUED TO
	93586

Marshall

93586

CPSIA information can be obtained
at www.ICGtesting.com
Printed in the USA
LVHW080309090822
725492LV00003B/39